ARCHITECTS CONTRACTORS ENGINEERS

DCD Guide To Construction COSTS

2019 Vol. L

Architects Contractors Engineers DCD Guide to Construction Costs 2019, Volume 50.

ISBN 978-1-58855-186-3

EDITOR'S NOTE 2019

This annually published book is designed to give a uniform estimating and cost control system to the General Building Contractor. It contains a complete system to be used with or without computers. It also contains Quick Estimating sections for preliminary conceptual budget estimates by Architects, Engineers and Contractors. Square Foot Estimating is also included for preliminary estimates.

The Metropolitan Area concept is also used and gives the cost modifiers to use for the variations between Metropolitan Areas. This encompasses over 80% of the industry. This book is published annually to be historically accurate with the traditional May-July wage contract settlements and to be a true construction year estimating and cost guide.

The Rate of Inflation in the Construction Industry in 2017 was 3.5%. Labor contributed a 2.3% increase and materials 4.6%.

The Wage Rate for Skilled Trades increased an average of 2.3% in 2017. Wage rates will probably increase at a 3% average next year.

The Material Rate was up in 2018. There were several significant changes. Price increases are rare with some noticeable decreases. Lumber is up considerably, and should be watched this year.

Construction Volume should be up again for 2019. Housing and Commercial Construction spending will be up. Highway and Heavy Construction will be coming back and governments seem to be recovering and there is talk still of improving our infrastructure.

The Construction Industry has had low to moderate inflation but that seems to be changing. Some materials should inflate at a slower pace, and some watched carefully.

We are recommending using a 5% increase in your estimates for work beyond July 1, 2018.

CONTENTS

Metropolitan Cost Index

The costs as presented in this book attempt to represent national averages. Costs, however, vary among regions, states and even between adjacent localities.

In order to more closely approximate the probable costs for specific locations throughout the U.S., this table of Metropolitan Cost Modifiers is provided in the following few pages. These adjustment factors are used to modify costs obtained from this book to help account for regional variations of construction costs and to provide a more accurate estimate for specific areas. The factors are formulated by comparing costs in a specific area to the costs as presented in the Costbook pages. An example of how to use these factors is shown below. Whenever local current costs are known, whether material prices or labor rates, they should be used when more accuracy is required.

| **Cost Obtained from Costbook Pages** | X | **Metroploitan Cost Multiplier Divided by 100** | = | **Adjusted Cost** |

For example, a project estimated to cost $125,000 using the Costbook pages can be adjusted to more closely approximate the cost in Los Angeles:

$$\$125,000 \quad X \quad \frac{119}{100} \quad = \quad \$148,750$$

1

State	Metropolitan Area	Multiplier
AK	ANCHORAGE	132
AL	ANNISTON	81
	AUBURN	82
	BIRMINGHAM	82
	DECATUR	84
	DOTHAN	83
	FLORENCE	84
	GADSDEN	82
	HUNTSVILLE	84
	MOBILE	86
	MONTGOMERY	81
	OPELIKA	82
	TUSCALOOSA	81
AR	FAYETTEVILLE	79
	FORT SMITH	79
	JONESBORO	78
	LITTLE ROCK	82
	NORTH LITTLE ROCK	82
	PINE BLUFF	80
	ROGERS	79
	SPRINGDALE	79
	TEXARKANA	79
AZ	FLAGSTAFF	94
	MESA	94
	PHOENIX	95
	TUCSON	93
	YUMA	94
CA	BAKERSFIELD	116
	CHICO	118
	FAIRFIELD	120
	FRESNO	118
	LODI	117
	LONG BEACH	119
	LOS ANGELES	119
	MERCED	118
	MODESTO	114
	NAPA	120
	OAKLAND	124
	ORANGE COUNTY	118
	PARADISE	114
	PORTERVILLE	116
	REDDING	114
	RIVERSIDE	116
	SACRAMENTO	118
	SALINAS	120
	SAN BERNARDINO	116
	SAN DIEGO	117
	SAN FRANCISCO	129
	SAN JOSE	126
	SAN LUIS OBISPO	113
	SANTA BARBARA	116
	SANTA CRUZ	120
	SANTA ROSA	121
	STOCKTON	117
	TULARE	118
	VALLEJO	120
	VENTURA	116
	VISALIA	118
	WATSONVILLE	118
	YOLO	118
	YUBA CITY	118
CO	BOULDER	103
	COLORADO SPRINGS	100
	DENVER	101
	FORT COLLINS	110
	GRAND JUNCTION	99
	GREELEY	108
	LONGMONT	103
	LOVELAND	110
	PUEBLO	105
CT	BRIDGEPORT	113
	DANBURY	113
	HARTFORD	112
	MERIDEN	113
	NEW HAVEN	113
	NEW LONDON	110
	NORWALK	117

State	Metropolitan Area	Multiplier
CT	NORWICH	110
	STAMFORD	117
	WATERBURY	112
DC	WASHINGTON	105
DE	DOVER	105
	NEWARK	106
	WILMINGTON	106
FL	BOCA RATON	80
	BRADENTON	80
	CAPE CORAL	78
	CLEARWATER	81
	DAYTONA BEACH	75
	FORT LAUDERDALE	83
	FORT MYERS	78
	FORT PIERCE	81
	FORT WALTON BEACH	76
	GAINESVILLE	80
	JACKSONVILLE	78
	LAKELAND	78
	MELBOURNE	75
	MIAMI	81
	NAPLES	79
	OCALA	79
	ORLANDO	77
	PALM BAY	75
	PANAMA CITY	77
	PENSACOLA	76
	PORT ST. LUCIE	81
	PUNTA GORDA	78
	SARASOTA	80
	ST. PETERSBURG	80
	TALLAHASSEE	75
	TAMPA	80
	TITUSVILLE	75
	WEST PALM BEACH	80
	WINTER HAVEN	78
GA	ALBANY	86
	ATHENS	89
	ATLANTA	92
	AUGUSTA	86
	COLUMBUS	79
	MACON	83
	SAVANNAH	87
HI	HONOLULU	138
IA	CEDAR FALLS	91
	CEDAR RAPIDS	102
	DAVENPORT	106
	DES MOINES	104
	DUBUQUE	95
	IOWA CITY	97
	SIOUX CITY	91
	WATERLOO	91
ID	BOISE CITY	102
	POCATELLO	102
IL	BLOOMINGTON	113
	CHAMPAIGN	109
	CHICAGO	125
	DECATUR	107
	KANKAKEE	113
	NORMAL	113
	PEKIN	111
	PEORIA	111
	ROCKFORD	113
	SPRINGFIELD	108
	URBANA	109
IN	BLOOMINGTON	102
	ELKHART	96
	EVANSVILLE	99
	FORT WAYNE	100
	GARY	107
	GOSHEN	96
	INDIANAPOLIS	103
	KOKOMO	101
	LAFAYETTE	101
	MUNCIE	101
	SOUTH BEND	102
	TERRE HAUTE	100

State	Metropolitan Area	Multiplier	State	Metropolitan Area	Multiplier
KS	KANSAS CITY	120	NC	HICKORY	72
	LAWRENCE	109		HIGH POINT	81
	TOPEKA	96		JACKSONVILLE	72
	WICHITA	87		LENOIR	72
KY	LEXINGTON	91		MORGANTON	72
	LOUISVILLE	102		RALEIGH	80
	OWENSBORO	101		ROCKY MOUNT	72
LA	ALEXANDRIA	89		WILMINGTON	72
	BATON ROUGE	93		WINSTON SALEM	77
	BOSSIER CITY	90	ND	BISMARCK	84
	HOUMA	93		FARGO	98
	LAFAYETTE	91		GRAND FORKS	81
	LAKE CHARLES	93	NE	LINCOLN	84
	MONROE	89		OMAHA	91
	NEW ORLEANS	95	NH	MANCHESTER	106
	SHREVEPORT	90		NASHUA	106
MA	BARNSTABLE	124		PORTSMOUTH	111
	BOSTON	128	NJ	ATLANTIC CITY	126
	BROCKTON	118		BERGEN	129
	FITCHBURG	120		BRIDGETON	125
	LAWRENCE	121		CAPE MAY	125
	LEOMINSTER	120		HUNTERDON	128
	LOWELL	124		JERSEY CITY	130
	NEW BEDFORD	118		MIDDLESEX	129
	PITTSFIELD	118		MILLVILLE	125
	SPRINGFIELD	119		MONMOUTH	129
	WORCESTER	120		NEWARK	129
	YARMOUTH	124		OCEAN	130
MD	BALTIMORE	95		PASSAIC	130
	CUMBERLAND	98		SOMERSET	128
	HAGERSTOWN	90		TRENTON	128
ME	AUBURN	87		VINELAND	125
	BANGOR	87	NM	ALBUQUERQUE	91
	LEWISTON	87		LAS CRUCES	91
	PORTLAND	88		SANTA FE	91
MI	ANN ARBOR	119	NV	LAS VEGAS	109
	BATTLE CREEK	111		RENO	97
	BAY CITY	116	NY	ALBANY	119
	BENTON HARBOR	111		BINGHAMTON	116
	DETROIT	120		BUFFALO	118
	EAST LANSING	117		DUTCHESS COUNTY	119
	FLINT	116		ELMIRA	118
	GRAND RAPIDS	112		GLENS FALLS	120
	HOLLAND	112		JAMESTOWN	112
	JACKSON	107		NASSAU	137
	KALAMAZOO	111		NEW YORK	148
	LANSING	117		NEWBURGH	119
	MIDLAND	115		NIAGARA FALLS	121
	MUSKEGON	112		ROCHESTER	118
	SAGINAW	116		ROME	109
MN	DULUTH	107		SCHENECTADY	119
	MINNEAPOLIS	112		SUFFOLK	137
	ROCHESTER	107		SYRACUSE	118
	ST. CLOUD	105		TROY	119
	ST. PAUL	112		UTICA	109
MO	COLUMBIA	114	OH	AKRON	112
	JOPLIN	103		CANTON	107
	KANSAS CITY	118		CINCINNATI	105
	SPRINGFIELD	96		CLEVELAND	114
	ST. JOSEPH	117		COLUMBUS	115
	ST. LOUIS	115		DAYTON	115
MS	BILOXI	79		ELYRIA	114
	GULFPORT	79		HAMILTON	105
	HATTIESBURG	79		LIMA	115
	JACKSON	79		LORAIN	114
	PASCAGOULA	79		MANSFIELD	115
MT	BILLINGS	96		MASSILLON	107
	GREAT FALLS	90		MIDDLETOWN	115
	MISSOULA	91		SPRINGFIELD	109
NC	ASHEVILLE	73		STEUBENVILLE	115
	CHAPEL HILL	79		TOLEDO	109
	CHARLOTTE	82		WARREN	111
	DURHAM	81		YOUNGSTOWN	111
	FAYETTEVILLE	75	OK	ENID	86
	GOLDSBORO	80		LAWTON	86
	GREENSBORO	81		OKLAHOMA CITY	85
	GREENVILLE	79		TULSA	80

3

State	Metropolitan Area	Multiplier
OR	ASHLAND	109
	CORVALLIS	112
	EUGENE	112
	MEDFORD	109
	PORTLAND	114
	SALEM	112
	SPRINGFIELD	112
PA	ALLENTOWN	118
	ALTOONA	110
	BETHLEHEM	118
	CARLISLE	113
	EASTON	118
	ERIE	112
	HARRISBURG	113
	HAZLETON	118
	JOHNSTOWN	104
	LANCASTER	113
	LEBANON	115
	PHILADELPHIA	134
	PITTSBURGH	116
	READING	119
	SCRANTON	116
	SHARON	112
	STATE COLLEGE	98
	WILKES BARRE	116
	WILLIAMSPORT	97
	YORK	113
PR	MAYAGUEZ	73
	PONCE	74
	SAN JUAN	75
RI	PROVIDENCE	122
SC	AIKEN	89
	ANDERSON	71
	CHARLESTON	76
	COLUMBIA	76
	FLORENCE	73
	GREENVILLE	76
	MYRTLE BEACH	73
	NORTH CHARLESTON	81
	SPARTANBURG	73
	SUMTER	76
SD	RAPID CITY	81
	SIOUX FALLS	85
TN	CHATTANOOGA	84
	CLARKSVILLE	83
	JACKSON	83
	JOHNSON CITY	83
	KNOXVILLE	80
	MEMPHIS	84
	NASHVILLE	83
TX	ABILENE	88
	AMARILLO	92
	ARLINGTON	87
	AUSTIN	89
	BEAUMONT	88
	BRAZORIA	88
	BROWNSVILLE	73
	BRYAN	86
	COLLEGE STATION	86
	CORPUS CHRISTI	84
	DALLAS	89
	DENISON	87
	EDINBURG	73
	EL PASO	81
	FORT WORTH	87
	GALVESTON	93
	HARLINGEN	73
	HOUSTON	88

State	Metropolitan Area	Multiplier
TX	KILLEEN	77
	LAREDO	78
	LONGVIEW	78
	LUBBOCK	91
	MARSHALL	87
	MCALLEN	73
	MIDLAND	87
	MISSION	73
	ODESSA	87
	PORT ARTHUR	88
	SAN ANGELO	87
	SAN ANTONIO	90
	SAN BENITO	73
	SAN MARCOS	89
	SHERMAN	87
	TEMPLE	77
	TEXARKANA	79
	TEXAS CITY	93
	TYLER	84
	VICTORIA	74
	WACO	77
	WICHITA FALLS	87
UT	OGDEN	95
	OREM	93
	PROVO	93
	SALT LAKE CITY	92
VA	CHARLOTTESVILLE	86
	LYNCHBURG	83
	NEWPORT NEWS	88
	NORFOLK	91
	PETERSBURG	78
	RICHMOND	90
	ROANOKE	76
	VIRGINIA BEACH	91
VT	BURLINGTON	97
WA	BELLEVUE	119
	BELLINGHAM	111
	BREMERTON	113
	EVERETT	117
	KENNEWICK	101
	OLYMPIA	113
	PASCO	100
	RICHLAND	101
	SEATTLE	119
	SPOKANE	98
	TACOMA	116
	YAKIMA	104
WI	APPLETON	113
	BELOIT	117
	EAU CLAIRE	113
	GREEN BAY	112
	JANESVILLE	117
	KENOSHA	118
	LA CROSSE	114
	MADISON	116
	MILWAUKEE	118
	NEENAH	113
	OSHKOSH	113
	RACINE	118
	SHEBOYGAN	112
	WAUKESHA	118
	WAUSAU	113
WV	CHARLESTON	113
	HUNTINGTON	113
	PARKERSBURG	113
	WHEELING	113
WY	CASPER	85
	CHEYENNE	85

HOW TO USE THIS BOOK

Labor Columns

> ➤ Units *include* Workers Comp., Unemployment, and FICA on labor (approx. 35%).
> ➤ Units *do not include* general conditions and equipment (approx. 10%).
> ➤ Units *do not include* contractors' overhead and profit (approx. 10%).
> ➤ Units are Government Prevailing wages.

Material Columns

> ➤ Units *do not include* general conditions and equipment (approx. 10%).
> ➤ Units *do not include* sales or use taxes (approx. 5%) of material cost.
> ➤ Units *do not include* contractors' overhead and profit (approx. 10%) and are FOB job site.

Total Columns (Subcontractors)

> ➤ Units *do not include* general contractors' overhead or profit (approx. 10%).

Quick Estimating Sections - For Preliminary and Conceptual Estimating

> ➤ Includes all labor, material, general conditions, equipment, taxes.

		UNIT	LABOR	MAT.	TOTAL
01020.10	**ALLOWANCES**				
0090	Overhead				
1000	$20,000 project				
1040	Average	PCT.			20.00
1080	$100,000 project				
1120	Average	PCT.			15.00
1160	$500,000 project				
1180	Average	PCT.			12.00
1220	$1,000,000 project				
1260	Average	PCT.			10.00
1480	Profit				
1500	$20,000 project				
1540	Average	PCT.			15.00
1580	$100,000 project				
1620	Average	PCT.			12.00
1660	$500,000 project				
1700	Average	PCT.			10.00
1740	$1,000,000 project				
1780	Average	PCT.			8.00
2000	Professional fees				
2100	Architectural				
2120	$100,000 project				
2160	Average	PCT.			10.00
2200	$500,000 project				
2240	Average	PCT.			8.00
2280	$1,000,000 project				
2320	Average	PCT.			7.00
4080	Taxes				
5000	Sales tax				
5040	Average	PCT.			5.00
5080	Unemployment				
5120	Average	PCT.			6.50
5200	Social security (FICA)	"			7.85
01050.10	**FIELD STAFF**				
1000	Superintendent				
1020	Minimum	YEAR			96,912
1040	Average	"			121,157
1060	Maximum	"			145,534
1080	Field engineer				
1100	Minimum	YEAR			95,586
1120	Average	"			109,711
1140	Maximum	"			126,088
1160	Foreman				
1180	Minimum	YEAR			64,414
1200	Average	"			103,014
1220	Maximum	"			120,609
1240	Bookkeeper/timekeeper				
1260	Minimum	YEAR			37,260
1280	Average	"			48,657
1300	Maximum	"			62,953
1320	Watchman				
1340	Minimum	YEAR			27,763

		UNIT	LABOR	MAT.	TOTAL
01050.10	**FIELD STAFF, Cont'd...**				
1360	Average	YEAR			37,138
1380	Maximum	"			46,880
01310.10	**SCHEDULING**				
0090	Scheduling for				
1000	$100,000 project				
1040	Average	PCT.			2.00
1080	$500,000 project				
1120	Average	PCT.			1.00
1160	$1,000,000 project				
1200	Average	PCT.			0.75
4000	Scheduling software				
4020	Minimum	EA.			680
4040	Average	"			3,880
4060	Maximum	"			77,630
01410.10	**TESTING**				
1080	Testing concrete, per test				
1100	Minimum	EA.			23.00
1120	Average	"			38.25
1140	Maximum	"			77.00
1160	Soil, per test				
1180	Minimum	EA.			47.00
1200	Average	"			120
1220	Maximum	"			310
01500.10	**TEMPORARY FACILITIES**				
1000	Barricades, temporary				
1010	Highway				
1020	Concrete	L.F.	4.93	14.00	18.93
1040	Wood	"	1.97	4.81	6.78
1060	Steel	"	1.64	4.98	6.62
1080	Pedestrian barricades				
1100	Plywood	S.F.	1.64	4.28	5.92
1120	Chain link fence	"	1.64	3.63	5.27
2000	Trailers, general office type, per month				
2020	Minimum	EA.			230
2040	Average	"			390
2060	Maximum	"			770
2080	Crew change trailers, per month				
2100	Minimum	EA.			140
2120	Average	"			150
2140	Maximum	"			230
01505.10	**MOBILIZATION**				
1000	Equipment mobilization				
1020	Bulldozer				
1040	Minimum	EA.			210
1060	Average	"			440
1080	Maximum	"			730
1100	Backhoe/front-end loader				
1120	Minimum	EA.			130
1140	Average	"			220
1160	Maximum	"			480
1180	Crane, crawler type				

		UNIT	LABOR	MAT.	TOTAL
01505.10	**MOBILIZATION, Cont'd...**				
1200	Minimum	EA.			2,300
1220	Average	"			5,650
1240	Maximum	"			12,140
1260	Truck crane				
1280	Minimum	EA.			530
1300	Average	"			810
1320	Maximum	"			1,400
1340	Pile driving rig				
1360	Minimum	EA.			10,460
1380	Average	"			20,930
1400	Maximum	"			37,680
01525.10	**CONSTRUCTION AIDS**				
1000	Scaffolding/staging, rent per month				
1020	Measured by lineal feet of base				
1040	10' high	L.F.			13.75
1060	20' high	"			25.00
1080	30' high	"			34.75
1100	40' high	"			40.25
1120	50' high	"			47.75
1140	Measured by square foot of surface				
1160	Minimum	S.F.			0.60
1180	Average	"			1.04
1200	Maximum	"			1.87
1220	Safety nets, heavy duty, per job				
1240	Minimum	S.F.			0.40
1260	Average	"			0.49
1280	Maximum	"			1.07
1300	Tarpaulins, fabric, per job				
1320	Minimum	S.F.			0.28
1340	Average	"			0.48
1360	Maximum	"			1.24
01525.20	**TEMPORARY CONST. SHELTERS**				
0010	Standard, alum. with fabric, 12'x20'x15' Ht.	S.F.			19.00
0020	12'x20'x20' Ht.	"			21.75
01570.10	**SIGNS**				
0080	Construction signs, temporary				
1000	Signs, 2' x 4'				
1020	Minimum	EA.			39.25
1040	Average	"			94.00
1060	Maximum	"			330
1160	Signs, 8' x 8'				
1180	Minimum	EA.			110
1200	Average	"			330
1220	Maximum	"			3,340
01600.10	**EQUIPMENT**				
0080	Air compressor				
1000	60 cfm				
1020	By day	EA.			100
1030	By week	"			300
1040	By month	"			920
1200	600 cfm				

01600.10	EQUIPMENT, Cont'd...	UNIT	LABOR	MAT.	TOTAL
1210	By day	EA.			590
1220	By week	"			1,750
1230	By month	"			5,300
1300	Air tools, per compressor, per day				
1310	Minimum	EA.			42.00
1320	Average	"			53.00
1330	Maximum	"			73.00
1400	Generators, 5 kw				
1410	By day	EA.			100
1420	By week	"			310
1430	By month	"			960
1500	Heaters, salamander type, per week				
1510	Minimum	EA.			130
1520	Average	"			180
1530	Maximum	"			380
1600	Pumps, submersible				
1605	50 gpm				
1610	By day	EA.			84.00
1620	By week	"			250
1630	By month	"			750
1675	500 gpm				
1680	By day	EA.			170
1690	By week	"			500
1700	By month	"			1,510
1900	Diaphragm pump, by week				
1920	Minimum	EA.			150
1930	Average	"			250
1940	Maximum	"			520
2000	Pickup truck				
2020	By day	EA.			160
2030	By week	"			460
2040	By month	"			1,430
2080	Dump truck				
2100	6 cy truck				
2120	By day	EA.			420
2130	By week	"			1,260
2140	By month	"			3,780
2300	16 cy truck				
2310	By day	EA.			840
2320	By week	"			2,520
2340	By month	"			7,560
2400	Backhoe, track mounted				
2420	1/2 cy capacity				
2430	By day	EA.			860
2440	By week	"			2,620
2450	By month	"			7,770
2500	1 cy capacity				
2510	By day	EA.			1,360
2520	By week	"			4,090
2530	By month	"			12,280
2600	3 cy capacity				
2620	By day	EA.			4,410

		UNIT	LABOR	MAT.	TOTAL
01600.10	**EQUIPMENT, Cont'd...**				
2640	By week	EA.			13,230
2680	By month	"			39,680
3000	Backhoe/loader, rubber tired				
3005	1/2 cy capacity				
3010	By day	EA.			520
3020	By week	"			1,570
3030	By month	"			4,720
3035	3/4 cy capacity				
3040	By day	EA.			630
3050	By week	"			1,890
3060	By month	"			5,670
3200	Bulldozer				
3205	75 hp				
3210	By day	EA.			730
3220	By week	"			2,200
3230	By month	"			6,610
4000	Cranes, crawler type				
4005	15 ton capacity				
4010	By day	EA.			940
4020	By week	"			2,830
4030	By month	"			8,500
4070	50 ton capacity				
4080	By day	EA.			2,100
4090	By week	"			6,300
4100	By month	"			18,900
4145	Truck mounted, hydraulic				
4150	15 ton capacity				
4160	By day	EA.			890
4170	By week	"			2,670
4180	By month	"			7,720
5380	Loader, rubber tired				
5385	1 cy capacity				
5390	By day	EA.			630
5400	By week	"			1,890
5410	By month	"			5,670
7000	Scraper				
7010	Elevated scraper, not including bulldozer, 12 c.y.				
7020	By day	EA.			1,440
7030	By week	"			4,310
7040	By month	"			12,240
7100	Self-propelled scraper, 14 c.y.				
7110	By day	EA.			2,880
7120	By week	"			8,640
7130	By month	"			24,400
01740.10	**BONDS**				
1000	Performance bonds				
1020	Minimum	PCT.			0.62
1040	Average	"			1.93
1060	Maximum	"			3.07

TABLE OF CONTENTS   PAGE

		UNIT	LABOR	MAT.	TOTAL
02210.10	**SOIL BORING**				
1000	Borings, uncased, stable earth				
1020	2-1/2" dia.	L.F.	34.00		34.00
1040	4" dia.	"	39.00		39.00
1500	Cased, including samples				
1520	2-1/2" dia.	L.F.	45.50		45.50
1540	4" dia.	"	78.00		78.00
02220.10	**COMPLETE BUILDING DEMOLITION**				
0200	Wood frame	C.F.	0.39		0.39
0300	Concrete	"	0.58		0.58
0400	Steel frame	"	0.78		0.78
02220.15	**SELECTIVE BUILDING DEMOLITION**				
1000	Partition removal				
1100	Concrete block partitions				
1120	4" thick	S.F.	2.46		2.46
1140	8" thick	"	3.29		3.29
1160	12" thick	"	4.48		4.48
1200	Brick masonry partitions				
1220	4" thick	S.F.	2.46		2.46
1240	8" thick	"	3.08		3.08
1260	12" thick	"	4.11		4.11
1280	16" thick	"	6.17		6.17
1380	Cast in place concrete partitions				
1400	Unreinforced				
1421	6" thick	S.F.	18.25		18.25
1423	8" thick	"	19.50		19.50
1425	10" thick	"	22.75		22.75
1427	12" thick	"	27.25		27.25
1440	Reinforced				
1441	6" thick	S.F.	21.00		21.00
1443	8" thick	"	27.25		27.25
1445	10" thick	"	30.25		30.25
1447	12" thick	"	36.25		36.25
1500	Terra cotta				
1520	To 6" thick	S.F.	2.46		2.46
1700	Stud partitions				
1720	Metal or wood, with drywall both sides	S.F.	2.46		2.46
1740	Metal studs, both sides, lath and plaster	"	3.29		3.29
2000	Door and frame removal				
2020	Hollow metal in masonry wall				
2030	Single				
2040	2'6"x6'8"	EA.	62.00		62.00
2060	3'x7'	"	82.00		82.00
2070	Double				
2080	3'x7'	EA.	99.00		99.00
2085	4'x8'	"	99.00		99.00
2140	Wood in framed wall				
2150	Single				
2160	2'6"x6'8"	EA.	35.25		35.25
2180	3'x6'8"	"	41.25		41.25
2190	Double				
2200	2'6"x6'8"	EA.	49.25		49.25

		UNIT	LABOR	MAT.	TOTAL
02220.15	**SELECTIVE BUILDING DEMOLITION, Cont'd...**				
2220	3'x6'8"	EA.	55.00		55.00
2240	Remove for re-use				
2260	Hollow metal	EA.	120		120
2280	Wood	"	82.00		82.00
2300	Floor removal				
2340	Brick flooring	S.F.	1.97		1.97
2360	Ceramic or quarry tile	"	1.09		1.09
2380	Terrazzo	"	2.19		2.19
2400	Heavy wood	"	1.31		1.31
2420	Residential wood	"	1.41		1.41
2440	Resilient tile or linoleum	"	0.49		0.49
2500	Ceiling removal				
2520	Acoustical tile ceiling				
2540	Adhesive fastened	S.F.	0.49		0.49
2560	Furred and glued	"	0.41		0.41
2580	Suspended grid	"	0.30		0.30
2600	Drywall ceiling				
2620	Furred and nailed	S.F.	0.54		0.54
2640	Nailed to framing	"	0.49		0.49
2660	Plastered ceiling				
2680	Furred on framing	S.F.	1.23		1.23
2700	Suspended system	"	1.64		1.64
2800	Roofing removal				
2820	Steel frame				
2840	Corrugated metal roofing	S.F.	0.98		0.98
2860	Built-up roof on metal deck	"	1.64		1.64
2900	Wood frame				
2920	Built up roof on wood deck	S.F.	1.51		1.51
2940	Roof shingles	"	0.82		0.82
2960	Roof tiles	"	1.64		1.64
8900	Concrete frame	C.F.	3.29		3.29
8920	Concrete plank	S.F.	2.46		2.46
8940	Built-up roof on concrete	"	1.41		1.41
9200	Cut-outs				
9230	Concrete, elevated slabs, mesh reinforcing				
9240	Under 5 cf	C.F.	49.25		49.25
9260	Over 5 cf	"	41.25		41.25
9270	Bar reinforcing				
9280	Under 5 cf	C.F.	82.00		82.00
9290	Over 5 cf	"	62.00		62.00
9300	Window removal				
9301	Metal windows, trim included				
9302	2'x3'	EA.	49.25		49.25
9304	2'x4'	"	55.00		55.00
9306	2'x6'	"	62.00		62.00
9308	3'x4'	"	62.00		62.00
9310	3'x6'	"	71.00		71.00
9312	3'x8'	"	82.00		82.00
9314	4'x4'	"	82.00		82.00
9315	4'x6'	"	99.00		99.00
9316	4'x8'	"	120		120
9317	Wood windows, trim included				

		UNIT	LABOR	MAT.	TOTAL
02220.15	**SELECTIVE BUILDING DEMOLITION, Cont'd...**				
9318	2'x3'	EA.	27.50		27.50
9319	2'x4'	"	29.00		29.00
9320	2'x6'	"	30.75		30.75
9321	3'x4'	"	33.00		33.00
9322	3'x6'	"	35.25		35.25
9324	3'x8'	"	38.00		38.00
9325	6'x4'	"	41.25		41.25
9326	6'x6'	"	44.75		44.75
9327	6'x8'	"	49.25		49.25
9329	Walls, concrete, bar reinforcing				
9330	Small jobs	C.F.	33.00		33.00
9340	Large jobs	"	27.50		27.50
9360	Brick walls, not including toothing				
9390	4" thick	S.F.	2.46		2.46
9400	8" thick	"	3.08		3.08
9410	12" thick	"	4.11		4.11
9415	16" thick	"	6.17		6.17
9420	Concrete block walls, not including toothing				
9440	4" thick	S.F.	2.74		2.74
9450	6" thick	"	2.90		2.90
9460	8" thick	"	3.08		3.08
9465	10" thick	"	3.52		3.52
9470	12" thick	"	4.11		4.11
9500	Rubbish handling				
9519	Load in dumpster or truck				
9520	Minimum	C.F.	1.09		1.09
9540	Maximum	"	1.64		1.64
9550	For use of elevators, add				
9560	Minimum	C.F.	0.24		0.24
9570	Maximum	"	0.49		0.49
9600	Rubbish hauling				
9640	Hand loaded on trucks, 2 mile trip	C.Y.	42.00		42.00
9660	Machine loaded on trucks, 2 mile trip	"	27.25		27.25
02225.20	**FENCE DEMOLITION**				
0060	Remove fencing				
0080	Chain link, 8' high				
0100	For disposal	L.F.	2.46		2.46
0200	For reuse	"	6.17		6.17
0980	Wood				
1000	4' high	S.F.	1.64		1.64
1960	Masonry				
1980	8" thick				
2000	4' high	S.F.	4.93		4.93
2020	6' high	"	6.17		6.17
02225.50	**SAW CUTTING PAVEMENT**				
0100	Pavement, bituminous				
0110	2" thick	L.F.	2.09		2.09
0120	3" thick	"	2.62		2.62
0200	Concrete pavement, with wire mesh				
0210	4" thick	L.F.	4.03		4.03
0212	5" thick	"	4.37		4.37

		UNIT	LABOR	MAT.	TOTAL
02225.50	**SAW CUTTING PAVEMENT, Cont'd...**				
0300	Plain concrete, unreinforced				
0320	4" thick	L.F.	3.49		3.49
0340	5" thick	"	4.03		4.03
02225.60	**TORCH CUTTING**				
0010	Steel plate, 1/2" thick,	L.F.	1.08		1.08
0020	1" thick	"	2.17		2.17
02230.50	**TREE CUTTING & CLEARING**				
0980	Cut trees and clear out stumps				
1000	9" to 12" dia.	EA.	540		540
1400	To 24" dia.	"	680		680
1600	24" dia. and up	"	910		910
02315.10	**BASE COURSE**				
1019	Base course, crushed stone				
1020	3" thick	S.Y.	0.74	3.50	4.24
1030	4" thick	"	0.80	4.71	5.51
1040	6" thick	"	0.87	7.07	7.94
2500	Base course, bank run gravel				
3020	4" deep	S.Y.	0.78	3.32	4.10
3040	6" deep	"	0.84	5.08	5.92
4000	Prepare and roll sub base				
4020	Minimum	S.Y.	0.74		0.74
4030	Average	"	0.92		0.92
4040	Maximum	"	1.23		1.23
02315.20	**BORROW**				
1000	Borrow fill, F.O.B. at pit				
1005	Sand, haul to site, round trip				
1010	10 mile	C.Y.	14.75	22.75	37.50
1020	20 mile	"	24.75	22.75	47.50
1030	30 mile	"	37.00	22.75	59.75
3980	Place borrow fill and compact				
4000	Less than 1 in 4 slope	C.Y.	7.41	22.75	30.16
4100	Greater than 1 in 4 slope	"	9.88	22.75	32.63
02315.30	**BULK EXCAVATION**				
1000	Excavation, by small dozer				
1020	Large areas	C.Y.	2.09		2.09
1040	Small areas	"	3.49		3.49
1060	Trim banks	"	5.24		5.24
1700	Hydraulic excavator				
1720	1 cy capacity				
1740	Light material	C.Y.	4.53		4.53
1760	Medium material	"	5.44		5.44
1780	Wet material	"	6.80		6.80
1790	Blasted rock	"	7.78		7.78
1800	1-1/2 cy capacity				
1820	Light material	C.Y.	1.85		1.85
1840	Medium material	"	2.47		2.47
1860	Wet material	"	2.96		2.96
2000	Wheel mounted front-end loader				
2020	7/8 cy capacity				
2040	Light material	C.Y.	3.70		3.70

		UNIT	LABOR	MAT.	TOTAL
02315.30	**BULK EXCAVATION, Cont'd...**				
2060	Medium material	C.Y.	4.23		4.23
2080	Wet material	"	4.94		4.94
2100	Blasted rock	"	5.92		5.92
2300	2-1/2 cy capacity				
2320	Light material	C.Y.	1.74		1.74
2340	Medium material	"	1.85		1.85
2360	Wet material	"	1.97		1.97
2380	Blasted rock	"	2.11		2.11
2600	Track mounted front-end loader				
2620	1-1/2 cy capacity				
2640	Light material	C.Y.	2.47		2.47
2660	Medium material	"	2.69		2.69
2680	Wet material	"	2.96		2.96
2700	Blasted rock	"	3.29		3.29
2720	2-3/4 cy capacity				
2740	Light material	C.Y.	1.48		1.48
2760	Medium material	"	1.64		1.64
2780	Wet material	"	1.85		1.85
2790	Blasted rock	"	2.11		2.11
4000	Scraper -500' haul				
4010	Elevated scraper, not including bulldozer, 12 c.y.				
4020	Light material	C.Y.	4.94		4.94
4030	Medium material	"	5.38		5.38
4040	Wet material	"	5.92		5.92
4050	Blasted rock	"	6.58		6.58
4100	Self-propelled scraper, 14 c.y.				
4110	Light material	C.Y.	4.56		4.56
4120	Medium material	"	4.94		4.94
4130	Wet material	"	5.38		5.38
4140	Blasted rock	"	5.92		5.92
4200	1,000' haul				
4210	Elevated scraper, not including bulldozer, 12 c.y.				
4220	Light material	C.Y.	5.92		5.92
4230	Medium material	"	6.58		6.58
4240	Wet material	"	7.41		7.41
4250	Blasted rock	"	8.46		8.46
4300	Self-propelled scraper, 14 c.y.				
4310	Light material	C.Y.	5.38		5.38
4320	Medium material	"	5.92		5.92
4330	Wet material	"	6.58		6.58
4340	Blasted rock	"	7.41		7.41
4400	2,000' haul				
4410	Elevated scraper, not including bulldozer, 12 c.y.				
4420	Light material	C.Y.	7.41		7.41
4430	Medium material	"	8.46		8.46
4440	Wet material	"	9.88		9.88
4450	Blasted rock	"	11.75		11.75
4510	Self-propelled scraper, 14 c.y.				
4520	Light material	C.Y.	6.58		6.58
4530	Medium material	"	7.41		7.41
4540	Wet material	"	8.46		8.46
4550	Blasted rock	"	9.88		9.88

		UNIT	LABOR	MAT.	TOTAL
02315.40	**BUILDING EXCAVATION**				
0090	Structural excavation, unclassified earth				
0100	3/8 cy backhoe	C.Y.	19.75		19.75
0110	3/4 cy backhoe	"	14.75		14.75
0120	1 cy backhoe	"	12.25		12.25
0600	Foundation backfill and compaction by machine	"	29.75		29.75
02315.45	**HAND EXCAVATION**				
0980	Excavation				
1000	To 2' deep				
1020	Normal soil	C.Y.	55.00		55.00
1040	Sand and gravel	"	49.25		49.25
1060	Medium clay	"	62.00		62.00
1080	Heavy clay	"	71.00		71.00
1100	Loose rock	"	82.00		82.00
1200	To 6' deep				
1220	Normal soil	C.Y.	71.00		71.00
1240	Sand and gravel	"	62.00		62.00
1260	Medium clay	"	82.00		82.00
1280	Heavy clay	"	99.00		99.00
1300	Loose rock	"	120		120
2020	Backfilling foundation without compaction, 6" lifts	"	30.75		30.75
2200	Compaction of backfill around structures or in trench				
2220	By hand with air tamper	C.Y.	35.25		35.25
2240	By hand with vibrating plate tamper	"	33.00		33.00
2250	1 ton roller	"	52.00		52.00
5400	Miscellaneous hand labor				
5440	Trim slopes, sides of excavation	S.F.	0.08		0.08
5450	Trim bottom of excavation	"	0.09		0.09
5460	Excavation around obstructions and services	C.Y.	160		160
02315.50	**ROADWAY EXCAVATION**				
0100	Roadway excavation				
0110	1/4 mile haul	C.Y.	2.96		2.96
0120	2 mile haul	"	4.94		4.94
0130	5 mile haul	"	7.41		7.41
3000	Spread base course	"	3.70		3.70
3100	Roll and compact	"	4.94		4.94
02315.60	**TRENCHING**				
0100	Trenching and continuous footing excavation				
0980	By gradall				
1000	1 cy capacity				
1040	Medium soil	C.Y.	4.56		4.56
1080	Loose rock	"	5.38		5.38
1090	Blasted rock	"	5.70		5.70
1095	By hydraulic excavator				
1100	1/2 cy capacity				
1140	Medium soil	C.Y.	5.38		5.38
1180	Loose rock	"	6.58		6.58
1190	Blasted rock	"	7.41		7.41
1200	1 cy capacity				
1240	Medium soil	C.Y.	3.70		3.70
1280	Loose rock	"	4.23		4.23
1300	Blasted rock	"	4.56		4.56

		UNIT	LABOR	MAT.	TOTAL
02315.60	**TRENCHING, Cont'd...**				
1600	2 cy capacity				
1640	Medium soil	C.Y.	3.12		3.12
1680	Loose rock	"	3.48		3.48
1690	Blasted rock	"	3.70		3.70
3000	Hand excavation				
3100	Bulk, wheeled 100'				
3120	Normal soil	C.Y.	55.00		55.00
3140	Sand or gravel	"	49.25		49.25
3160	Medium clay	"	71.00		71.00
3180	Heavy clay	"	99.00		99.00
3200	Loose rock	"	120		120
3300	Trenches, up to 2' deep				
3320	Normal soil	C.Y.	62.00		62.00
3340	Sand or gravel	"	55.00		55.00
3360	Medium clay	"	82.00		82.00
3380	Heavy clay	"	120		120
3390	Loose rock	"	160		160
3400	Trenches, to 6' deep				
3420	Normal soil	C.Y.	71.00		71.00
3440	Sand or gravel	"	62.00		62.00
3460	Medium clay	"	99.00		99.00
3480	Heavy clay	"	160		160
3500	Loose rock	"	250		250
3590	Backfill trenches				
3600	With compaction				
3620	By hand	C.Y.	41.25		41.25
3640	By 60 hp tracked dozer	"	2.62		2.62
02315.70	**UTILITY EXCAVATION**				
2080	Trencher, sandy clay, 8" wide trench				
2100	18" deep	L.F.	2.33		2.33
2200	24" deep	"	2.62		2.62
2300	36" deep	"	2.99		2.99
6080	Trench backfill, 95% compaction				
7000	Tamp by hand	C.Y.	30.75		30.75
7050	Vibratory compaction	"	24.75		24.75
7060	Trench backfilling, with borrow sand, place & compact	"	24.75	22.75	47.50
02315.80	**HAULING MATERIAL**				
0090	Haul material by 10 cy dump truck, round trip distance				
0100	1 mile	C.Y.	5.83		5.83
0110	2 mile	"	6.99		6.99
0120	5 mile	"	9.54		9.54
0130	10 mile	"	10.50		10.50
0140	20 mile	"	11.75		11.75
0150	30 mile	"	14.00		14.00
02340.05	**SOIL STABILIZATION**				
0100	Straw bale secured with rebar	L.F.	1.64	7.54	9.18
0120	Filter barrier, 18" high filter fabric	"	4.93	1.82	6.75
0130	Sediment fence, 36" fabric with 6" mesh	"	6.17	4.32	10.49
1000	Soil stabilization with tar paper, burlap, straw and stakes	S.F.	0.07	0.36	0.43

		UNIT	LABOR	MAT.	TOTAL
02360.20	**SOIL TREATMENT**				
1100	Soil treatment, termite control pretreatment				
1120	Under slabs	S.F.	0.27	0.38	0.65
1140	By walls	"	0.32	0.38	0.70
02370.40	**RIPRAP**				
0100	Riprap				
0110	Crushed stone blanket, max size 2-1/2"	TON	78.00	35.25	113
0120	Stone, quarry run, 300 lb. stones	"	72.00	44.25	116
0130	400 lb. stones	"	67.00	46.00	113
0140	500 lb. stones	"	63.00	48.00	111
0150	750 lb. stones	"	59.00	49.75	109
0160	Dry concrete riprap in bags 3" thick, 80 lb. per bag	BAG	3.92	5.96	9.88
02455.60	**STEEL PILES**				
1000	H-section piles				
1010	8x8				
1020	36 lb/ft				
1021	30' long	L.F.	13.50	16.75	30.25
1022	40' long	"	10.75	16.75	27.50
5000	Tapered friction piles, fluted casing, up to 50'				
5002	With 4000 psi concrete no reinforcing				
5040	12" dia.	L.F.	8.14	19.50	27.64
5060	14" dia.	"	8.34	22.50	30.84
02455.65	**STEEL PIPE PILES**				
1000	Concrete filled, 3000# concrete, up to 40'				
1100	8" dia.	L.F.	11.50	23.00	34.50
1120	10" dia.	"	12.00	29.50	41.50
1140	12" dia.	"	12.50	34.25	46.75
2000	Pipe piles, non-filled				
2020	8" dia.	L.F.	9.04	21.75	30.79
2040	10" dia.	"	9.30	27.25	36.55
2060	12" dia.	"	9.57	33.25	42.82
2520	Splice				
2540	8" dia.	EA.	99.00	97.00	196
2560	10" dia.	"	99.00	110	209
2580	12" dia.	"	120	120	240
2680	Standard point				
2700	8" dia.	EA.	99.00	130	229
2740	10" dia.	"	99.00	170	269
2760	12" dia.	"	120	180	300
2880	Heavy duty point				
2900	8" dia.	EA.	120	230	350
2920	10" dia.	"	120	320	440
2940	12" dia.	"	160	340	500
02455.80	**WOOD AND TIMBER PILES**				
0080	Treated wood piles, 12" butt, 8" tip				
0100	25' long	L.F.	16.25	16.50	32.75
0110	30' long	"	13.50	17.50	31.00
0120	35' long	"	11.50	17.50	29.00
0125	40' long	"	10.25	17.50	27.75

		UNIT	LABOR	MAT.	TOTAL
02465.50	**PRESTRESSED PILING**				
0980	Prestressed concrete piling, less than 60' long				
1000	10" sq.	L.F.	6.78	20.75	27.53
1002	12" sq.	"	7.07	29.00	36.07
1480	Straight cylinder, less than 60' long				
1500	12" dia.	L.F.	7.40	27.00	34.40
1540	14" dia.	"	7.57	36.50	44.07
02510.10	**WELLS**				
0980	Domestic water, drilled and cased				
1000	4" dia.	L.F.	81.00	30.50	112
1020	6" dia.	"	90.00	33.50	124
02510.40	**DUCTILE IRON PIPE**				
0990	Ductile iron pipe, cement lined, slip-on joints				
1000	4"	L.F.	7.56	18.25	25.81
1010	6"	"	8.01	21.25	29.26
1020	8"	"	8.51	27.75	36.26
1190	Mechanical joint pipe				
1200	4"	L.F.	10.50	19.75	30.25
1210	6"	"	11.25	23.50	34.75
1220	8"	"	12.50	31.00	43.50
1480	Fittings, mechanical joint				
1500	90 degree elbow				
1520	4"	EA.	33.00	230	263
1540	6"	"	38.00	300	338
1560	8"	"	49.25	430	479
1700	45 degree elbow				
1720	4"	EA.	33.00	200	233
1740	6"	"	38.00	270	308
1760	8"	"	49.25	380	429
02510.60	**PLASTIC PIPE**				
0110	PVC, class 150 pipe				
0120	4" dia.	L.F.	6.80	5.31	12.11
0130	6" dia.	"	7.36	10.00	17.36
0140	8" dia.	"	7.78	16.00	23.78
0165	Schedule 40 pipe				
0170	1-1/2" dia.	L.F.	2.90	1.34	4.24
0180	2" dia.	"	3.08	1.99	5.07
0185	2-1/2" dia.	"	3.29	3.01	6.30
0190	3" dia.	"	3.52	4.09	7.61
0200	4" dia.	"	4.11	5.78	9.89
0210	6" dia.	"	4.93	11.00	15.93
0240	90 degree elbows				
0250	1"	EA.	8.22	1.12	9.34
0260	1-1/2"	"	8.22	2.14	10.36
0270	2"	"	8.97	3.35	12.32
0280	2-1/2"	"	9.87	10.25	20.12
0290	3"	"	11.00	12.25	23.25
0300	4"	"	12.25	19.75	32.00
0310	6"	"	16.50	62.00	78.50
0500	Couplings				
0510	1"	EA.	8.22	0.91	9.13
0520	1-1/2"	"	8.22	1.30	9.52

		UNIT	LABOR	MAT.	TOTAL
02510.60	**PLASTIC PIPE, Cont'd...**				
0530	2"	EA.	8.97	2.01	10.98
0540	2-1/2"	"	9.87	4.42	14.29
0550	3"	"	11.00	6.91	17.91
0560	4"	"	12.25	9.02	21.27
0580	6"	"	16.50	28.50	45.00
02530.20	**VITRIFIED CLAY PIPE**				
0100	Vitrified clay pipe, extra strength				
1020	6" dia.	L.F.	12.50	5.21	17.71
1040	8" dia.	"	13.00	6.24	19.24
1050	10" dia.	"	13.50	9.57	23.07
02530.30	**MANHOLES**				
0100	Precast sections, 48" dia.				
0110	Base section	EA.	230	360	590
0120	1'0" riser	"	180	100	280
0130	1'4" riser	"	190	120	310
0140	2'8" riser	"	210	180	390
0150	4'0" riser	"	230	340	570
0160	2'8" cone top	"	270	220	490
0170	Precast manholes, 48" dia.				
0180	4' deep	EA.	540	700	1,240
0200	6' deep	"	680	1,070	1,750
0250	7' deep	"	780	1,220	2,000
0260	8' deep	"	910	1,380	2,290
0280	10' deep	"	1,090	1,540	2,630
1000	Cast-in-place, 48" dia., with frame and cover				
1100	5' deep	EA.	1,360	630	1,990
1120	6' deep	"	1,560	830	2,390
1140	8' deep	"	1,820	1,210	3,030
1160	10' deep	"	2,180	1,410	3,590
1480	Brick manholes, 48" dia. with cover, 8" thick				
1500	4' deep	EA.	600	670	1,270
1501	6' deep	"	670	840	1,510
1505	8' deep	"	750	1,080	1,830
1510	10' deep	"	860	1,340	2,200
4200	Frames and covers, 24" diameter				
4210	300 lb	EA.	49.25	410	459
4220	400 lb	"	55.00	430	485
4980	Steps for manholes				
5000	7" x 9"	EA.	9.87	17.25	27.12
5020	8" x 9"	"	11.00	22.00	33.00
02530.40	**SANITARY SEWERS**				
0980	Clay				
1000	6" pipe	L.F.	9.07	8.55	17.62
2980	PVC				
3000	4" pipe	L.F.	6.80	3.68	10.48
3010	6" pipe	"	7.16	7.38	14.54
02540.10	**DRAINAGE FIELDS**				
0080	Perforated PVC pipe, for drain field				
0100	4" pipe	L.F.	6.05	2.46	8.51
0120	6" pipe	"	6.48	4.62	11.10

		UNIT	LABOR	MAT.	TOTAL
02540.50	**SEPTIC TANKS**				
0980	Septic tank, precast concrete				
1000	1000 gals	EA.	450	990	1,440
1200	2000 gals	"	680	2,660	3,340
1310	Leaching pit, precast concrete, 72" diameter				
1320	3' deep	EA.	340	760	1,100
1340	6' deep	"	390	1,330	1,720
1360	8' deep	"	450	1,690	2,140
02630.70	**UNDERDRAIN**				
1480	Drain tile, clay				
1500	6" pipe	L.F.	6.05	4.52	10.57
1520	8" pipe	"	6.33	7.21	13.54
1580	Porous concrete, standard strength				
1600	6" pipe	L.F.	6.05	4.68	10.73
1620	8" pipe	"	6.33	5.06	11.39
1800	Corrugated metal pipe, perforated type				
1810	6" pipe	L.F.	6.80	6.49	13.29
1820	8" pipe	"	7.16	7.66	14.82
1980	Perforated clay pipe				
2000	6" pipe	L.F.	7.78	5.44	13.22
2020	8" pipe	"	8.01	7.29	15.30
2480	Drain tile, concrete				
2500	6" pipe	L.F.	6.05	3.71	9.76
2520	8" pipe	"	6.33	5.77	12.10
4980	Perforated rigid PVC underdrain pipe				
5000	4" pipe	L.F.	4.53	1.91	6.44
5100	6" pipe	"	5.44	3.67	9.11
5150	8" pipe	"	6.05	5.61	11.66
6980	Underslab drainage, crushed stone				
7000	3" thick	S.F.	0.90	0.30	1.20
7120	4" thick	"	1.04	0.40	1.44
7140	6" thick	"	1.13	0.62	1.75
7180	Plastic filter fabric for drain lines	"	0.49	0.46	0.95
02740.20	**ASPHALT SURFACES**				
0050	Asphalt wearing surface, flexible pavement				
0100	1" thick	S.Y.	2.71	4.52	7.23
0120	1-1/2" thick	"	3.25	6.82	10.07
1000	Binder course				
1010	1-1/2" thick	S.Y.	3.01	6.45	9.46
1030	2" thick	"	3.70	8.58	12.28
2000	Bituminous sidewalk, no base				
2020	2" thick	S.Y.	3.20	9.84	13.04
2040	3" thick	"	3.40	14.75	18.15
02750.10	**CONCRETE PAVING**				
1080	Concrete paving, reinforced, 5000 psi concrete				
2000	6" thick	S.Y.	25.50	25.00	50.50
2005	7" thick	"	27.25	29.25	56.50
2010	8" thick	"	29.00	33.25	62.25
02810.40	**LAWN IRRIGATION**				
0480	Residential system, complete				
0500	Minimum	ACRE			16,740
0520	Maximum	"			31,860

		UNIT	LABOR	MAT.	TOTAL
02820.10	**CHAIN LINK FENCE**				
0230	Chain link fence, 9 ga., galvanized, with posts 10' o.c.				
0250	4' high	L.F.	3.52	7.38	10.90
0260	5' high	"	4.48	9.87	14.35
0270	6' high	"	6.17	11.25	17.42
1161	Fabric, galvanized chain link, 2" mesh, 9 ga.				
1163	4' high	L.F.	1.64	4.01	5.65
1164	5' high	"	1.97	4.91	6.88
1165	6' high	"	2.46	6.87	9.33
02820.20	**WOOD FENCE**				
0010	4X4 posts w/2x4 horizontals, 4'-8' high				
0020	Cedar, 1x4 planks, picket	S.F.	1.23	1.93	3.16
0030	1x6 planks, privacy	"	0.98	3.06	4.04
0040	1x8 planks, privacy	"	0.89	2.80	3.69
0050	Redwood, 1x4 planks, picket	"	1.23	1.57	2.80
0060	1x6 planks, privacy	"	0.98	2.91	3.89
0070	1x8 planks	"	0.89	2.54	3.43
0080	Treated pine, 1x4 planks, picket	"	1.23	1.55	2.78
0100	1x6 planks, privacy	"	0.98	1.51	2.49
0110	1x8 planks, privacy	"	0.89	1.48	2.37
0210	4' high				
0220	Composite, 1x4 planks, picket	S.F.	1.54	1.69	3.23
0230	1x6 planks, privacy	"	1.23	2.91	4.14
0240	1x8 planks	"	1.02	2.54	3.56
0250	Vinyl, 1x4 planks, picket	"	1.54	1.51	3.05
0260	1x6 planks, privacy	"	1.23	2.68	3.91
0270	1x8 planks, privacy	"	1.02	2.64	3.66
0280	Gate, cedar or redwood, 3' w, 1x4 planks, picket	EA.	41.25	26.50	67.75
0290	1x6 planks, privacy	"	49.25	73.00	122
0300	1x8 planks, privacy	"	55.00	83.00	138
02840.40	**PARKING BARRIERS**				
3010	Bollard, conc. filled, 8' long				
3020	6" dia.	EA.	42.00	540	582
3030	8" dia.	"	53.00	810	863
3040	12" dia.	"	63.00	1,040	1,103
02880.70	**RECREATIONAL COURTS**				
1000	Walls, galvanized steel				
1020	8' high	L.F.	9.87	16.00	25.87
1040	10' high	"	11.00	18.75	29.75
1060	12' high	"	13.00	21.75	34.75
1200	Vinyl coated				
1220	8' high	L.F.	9.87	15.25	25.12
1240	10' high	"	11.00	18.75	29.75
1260	12' high	"	13.00	20.75	33.75
2010	Gates, galvanized steel				
2200	Single, 3' transom				
2210	3'x7'	EA.	250	370	620
2220	4'x7'	"	280	390	670
2230	5'x7'	"	330	540	870
2240	6'x7'	"	390	580	970
2400	Vinyl coated				
2405	Single, 3' transom				

		UNIT	LABOR	MAT.	TOTAL
02880.70	**RECREATIONAL COURTS, Cont'd...**				
2410	3'x7'	EA.	250	730	980
2420	4'x7'	"	280	790	1,070
2430	5'x7'	"	330	790	1,120
2440	6'x7'	"	390	820	1,210
02910.10	**TOPSOIL**				
0005	Spread topsoil, with equipment				
0010	Minimum	C.Y.	14.75		14.75
0020	Maximum	"	18.50		18.50
0080	By hand				
0100	Minimum	C.Y.	49.25		49.25
0110	Maximum	"	62.00		62.00
0980	Area prep. seeding (grade, rake and clean)				
1000	Square yard	S.Y.	0.39		0.39
1020	By acre	ACRE	1,970		1,970
2000	Remove topsoil and stockpile on site				
2020	4" deep	C.Y.	12.25		12.25
2040	6" deep	"	11.50		11.50
2200	Spreading topsoil from stock pile				
2220	By loader	C.Y.	13.50		13.50
2240	By hand	"	150		150
2260	Top dress by hand	S.Y.	1.48		1.48
2280	Place imported top soil				
2300	By loader				
2320	4" deep	S.Y.	1.48		1.48
2340	6" deep	"	1.64		1.64
2360	By hand				
2370	4" deep	S.Y.	5.48		5.48
2380	6" deep	"	6.17		6.17
5980	Plant bed preparation, 18" deep				
6000	With backhoe/loader	S.Y.	3.70		3.70
6010	By hand	"	8.22		8.22
02920.10	**FERTILIZING**				
0080	Fertilizing (23#/1000 sf)				
0100	By square yard	S.Y.	0.16	0.03	0.19
0120	By acre	ACRE	820	170	990
2980	Liming (70#/1000 sf)				
3000	By square yard	S.Y.	0.21	0.03	0.24
3020	By acre	ACRE	1,090	170	1,260
02920.30	**SEEDING**				
0980	Mechanical seeding, 175 lb/acre				
1000	By square yard	S.Y.	0.13	0.20	0.33
1020	By acre	ACRE	650	810	1,460
2040	450 lb/acre				
2060	By square yard	S.Y.	0.16	0.51	0.67
2080	By acre	ACRE	820	2,010	2,830
5980	Seeding by hand, 10 lb per 100 s.y.				
6000	By square yard	S.Y.	0.16	0.57	0.73
6010	By acre	ACRE	820	2,240	3,060
8010	Reseed disturbed areas	S.F.	0.24	0.05	0.29

		UNIT	LABOR	MAT.	TOTAL
02935.10	**SHRUB & TREE MAINTENANCE**				
1000	Moving shrubs on site				
1220	3' high	EA.	49.25		49.25
1240	4' high	"	55.00		55.00
2000	Moving trees on site				
3060	6' high	EA.	61.00		61.00
3080	8' high	"	68.00		68.00
3100	10' high	"	91.00		91.00
3110	Palm trees				
3140	10' high	EA.	91.00		91.00
3144	40' high	"	540		540
02935.30	**WEED CONTROL**				
1000	Weed control, bromicil, 15 lb./acre, wettable powder	ACRE	250	310	560
1100	Vegetation control, by application of plant killer	S.Y.	0.19	0.02	0.21
1200	Weed killer, lawns and fields	"	0.09	0.26	0.35
02945.20	**LANDSCAPE ACCESSORIES**				
0100	Steel edging, 3/16" x 4"	L.F.	0.61	1.29	1.90
0200	Landscaping stepping stones, 15"x15", white	EA.	2.46	5.83	8.29
6000	Wood chip mulch	C.Y.	33.00	40.75	73.75
6010	2" thick	S.Y.	0.98	2.49	3.47
6020	4" thick	"	1.41	4.70	6.11
6030	6" thick	"	1.79	7.04	8.83
6200	Gravel mulch, 3/4" stone	C.Y.	49.25	32.25	81.50
6300	White marble chips, 1" deep	S.F.	0.49	0.63	1.12
6980	Peat moss				
7000	2" thick	S.Y.	1.09	3.46	4.55
7020	4" thick	"	1.64	6.66	8.30
7030	6" thick	"	2.05	10.25	12.30
7980	Landscaping timbers, treated lumber				
8000	4" x 4"	L.F.	1.64	2.02	3.66
8020	6" x 6"	"	1.76	4.04	5.80
8040	8" x 8"	"	2.05	6.60	8.65

		UNIT	COST
02999.10	**DEMOLITION**		
3200	Selective Building Removals		
3210	No Cutting or Disposal Included		
3220	Concrete - Hand Work		
3230	8" Walls - Reinforced	S.F.	17.00
3240	Non - Reinforced	"	11.25
3250	12" Walls - Reinforced	"	27.25
3260	12" Footings x 24" wide	L.F.	24.75
3270	x 36" wide	"	34.00
3280	16" Footings x 24" wide	S.F.	45.50
3290	6" Structural Slab - Reinforced	"	10.50
3300	8" Structural Slab - Reinforced	"	13.00
3310	4" Slab on Ground - Reinforced	"	4.86
3320	Non - Reinforced	"	3.63
3330	6" Slab on Ground - Reinforced	"	5.79
3340	Non - Reinforced	"	4.32
3350	Stairs - Reinforced	"	18.25
3360	Masonry - Hand Work		
3370	4" Brick or Stone Walls	S.F.	3.28
3380	4" Brick and 8" Backup Block or Tile	"	6.05
3390	4" Block or Tile Partitions	"	3.20
3400	6" Block or Tile Partitions	"	3.49
3410	8" Block or Tile Partitions	"	4.00
3420	12" Block or Tile Partitions	"	5.04
3430	Miscellaneous - Hand Work		
3440	Acoustical Ceilings - Attached	S.F.	0.93
3450	Suspended (Including Grid)	"	0.53
3460	Asbestos - Pipe	L.F.	68.00
3470	Ceilings and Walls	S.F.	22.75
3480	Columns and Beams	"	54.00
3490	Tile Flooring	"	2.54
3500	Cabinets and Tops	L.F.	15.25
3510	Carpet	S.F.	0.46
3520	Ceramic and Quarry Tile	"	1.81
3530	Doors and Frames - Metal	EA.	80.00
3540	Wood	"	68.00
3550	Drywall Ceilings - Attached	S.F.	1.14
3560	Drywall on Wood or Metal Studs - 2 Sides	"	1.27
3570	Paint Removal - Doors and Windows	"	1.14
3580	Walls	"	0.93
3590	Plaster Ceilings - Attached (Including Iron)	"	1.61
3600	on Wood or Metal Studs	"	1.54
3610	Roofing - Builtup	"	1.54
3620	Shingles - Asphalt and Wood	"	0.53
3630	Terrazzo Flooring	"	2.55
3640	Vinyl Flooring	"	0.53
3650	Wall Coverings	"	1.00
3660	Windows	EA.	54.00
3670	Wood Flooring	S.F.	0.53
4300	Site Removals (Including Loading)		
4310	4" Concrete Walks - Labor Only (Non-Reinforced)	S.F.	3.63
4320	Machine Only (Non-Reinforced)	"	1.20
4330	6" Concrete Drives - Labor Only (Non-Reinforced)	"	4.53

		UNIT	COST
02999.10	**DEMOLITION, Cont'd...**		
4340	Machine Only (Reinforced)	S.F.	1.61
4350	6" x 18" Concrete Curb - Machine	L.F.	3.09
4360	Curb and Gutter - Machine	"	4.02
4370	2" Asphalt - Machine	S.F.	0.69
4380	Fencing - 8' Hand	L.F.	4.15
02999.20	**EARTHWORK**		
1000	GRADING - Hand - 4" - Site	S.F.	0.41
1010	Hand – 4" Building	"	0.69
2000	EXCAVATION - Hand - Open - Soft (Sand)	C.Y.	54.00
2010	Medium (Clay)	"	68.00
2020	Hard (Shale)	"	110
2030	Add for Trench or Pocket	PCT.	15.00
3000	BACK FILL - Hand - Not Comp. (Site Borrow)	C.Y.	24.75
4000	BACK FILL - Hand - Comp.(Site Borrow)		
4010	12" Lifts - Building - No Machine	C.Y.	42.00
4020	With Machine	"	34.00
4030	18" Lifts - Building - No Machine	"	39.00
4040	With Machine	"	32.00
02999.50	**DRAINAGE**		
4000	BUILDING FOUNDATION DRAINAGE		
4010	4" Clay Pipe	L.F.	10.13
4020	4" Plastic Pipe - Perforated	"	8.15
4030	6" Clay Pipe	"	11.45
4040	6" Plastic Pipe - Perforated	"	12.91
4050	Add for Porous Surround - 2' x 2'	"	14.90
02999.60	**PAVEMENT, CURBS AND WALKS**		
2000	CURBS AND GUTTERS		
2100	Concrete - Cast in Place (Machine Placed)		
2110	Curb - 6" x 12"	L.F.	24.22
2120	6" x 18"	"	28.29
2130	6" x 24"	"	32.49
2140	6" x 30"	"	37.06
2150	Curb and Gutter - 6" x 12"	"	34.93
2160	6" x 18"	"	41.25
2170	6" x 24"	"	47.25
2180	Add for Hand Placed	"	11.25
2190	Add for 2 #5 Reinf. Rods	"	5.02
2191	Add for Curves and Radius Work	PCT.	44.00
2200	Concrete Precast - 6" x 10" x 8"	L.F.	26.22
2210	6" x 9" x 8"	"	23.85
2250	6" x 8"	"	21.03
2300	Bituminous - 6" x 8"	"	9.27
2400	Granite - 6" x 16"	"	76.00
2500	Timbers - Treated - 6" x 6"	"	17.34
2600	Plastic - 6" x 6"	"	22.72
3000	WALKS		
3010	Bituminous - 1 1/2" with 4" Sand Base	S.F.	4.56
3020	2" with 4" Sand Base	"	5.02
3030	Concrete - 4" - Broom Finish	"	8.39
3040	5" - Broom Finish	"	9.50
3050	6" - Broom Finish	"	10.73

		UNIT	COST
02999.60	**PAVEMENT, CURBS AND WALKS, Cont'd...**		
3060	Add for 6" x 6", 10 - 10 Mesh	S.F.	1.90
3070	Add for 4" Sand Base	"	1.65
3080	Add for Exposed Aggregate	"	2.62
3090	Crushed Rock - 4"	"	1.79
4000	Brick - 4" - with 2" Sand Cushion	"	20.52
4010	4" - with 2" Mortar Setting Bed	"	24.54
4020	Flagstone – 1 1/4" - with 4" Sand Cushion	"	33.39
4030	1¼" - with 2" Mortar Setting Bed	"	37.50
4040	Precast Block - 1" Colored with 4" Sand Cushion	"	9.23
4050	2" Colored with 4" Sand Cushion	"	10.70
4060	Wood - 2" Boards on 6" x 6" Timbers	"	16.28
4070	2" Boards on 4" x 4" Timbers	"	13.77
4080	Slate - 1 1/4" - with 2" Mortar Setting Bed	"	44.75

TABLE OF CONTENTS   PAGE

		UNIT	LABOR	MAT.	TOTAL
03100.03	**FORMWORK ACCESSORIES**				
1000	Column clamps				
1010	Small, adjustable, 24"x24"	EA.			87.00
1020	Medium 36"x36"	"			90.00
1030	Large 60"x60"	"			91.00
2000	Forming hangers				
2010	Iron 14 ga.	EA.			2.81
2020	22 ga.	"			2.81
3000	Snap ties				
3010	Short-end with washers, 6" long	EA.			1.73
3020	12" long	"			1.96
4000	18" long	"			2.34
4010	24" long	"			2.52
4020	Long-end with washers, 6' long	"			2.04
4030	12" long	"			2.26
4040	18" long	"			2.56
4050	24" long	"			2.87
5000	Stakes				
5010	Round, pre-drilled holes, 12" long	EA.			6.14
5020	18" long	"			6.87
5030	24" long	"			8.90
5040	30" long	"			11.50
5050	36" long	"			13.75
5060	48" long	"			18.50
5070	I beam type, 12" long	"			5.05
6000	18" long	"			5.75
6010	24" long	"			8.73
6020	30" long	"			10.25
6030	36" long	"			13.00
7000	48" long	"			16.25
7010	Taper ties				
7020	50K, 1-1/4" to 1", 35" long	EA.			85.00
8000	45" long	"			140
8010	55" long	"			170
8020	Walers				
8030	5" deep, 4' long	EA.			240
8040	8' long	"			310
8050	12' long	"			510
9000	16' long	"			660
9010	8" deep, 4' long	"			320
9020	8' long	"			580
9030	12' long	"			840
9040	16' long	"			1,330
03110.05	**BEAM FORMWORK**				
1000	Beam forms, job built				
1020	Beam bottoms				
1040	1 use	S.F.	10.50	5.15	15.65
1080	3 uses	"	9.69	2.32	12.01
1120	5 uses	"	9.00	1.75	10.75
2000	Beam sides				
2020	1 use	S.F.	7.00	3.68	10.68
2060	3 uses	"	6.30	1.92	8.22

		UNIT	LABOR	MAT.	TOTAL
03110.05	**BEAM FORMWORK, Cont'd...**				
2100	5 uses	S.F.	5.72	1.56	7.28
03110.15	**COLUMN FORMWORK**				
1000	Column, square forms, job built				
1020	8" x 8" columns				
1040	1 use	S.F.	12.50	4.33	16.83
1080	3 uses	"	11.75	1.97	13.72
1120	5 uses	"	10.75	1.53	12.28
1300	16" x 16" columns				
1320	1 use	S.F.	10.50	3.77	14.27
1360	3 uses	"	9.84	1.59	11.43
1390	5 uses	"	9.26	1.19	10.45
2000	Round fiber forms, 1 use				
2040	10" dia.	L.F.	12.50	5.65	18.15
2120	18" dia.	"	15.00	19.50	34.50
2180	36" dia.	"	19.00	44.50	63.50
03110.18	**CURB FORMWORK**				
0980	Curb forms				
0990	Straight, 6" high				
1000	1 use	L.F.	6.30	2.58	8.88
1040	3 uses	"	5.72	1.16	6.88
1080	5 uses	"	5.25	0.94	6.19
1090	Curved, 6" high				
2000	1 use	L.F.	7.87	2.79	10.66
2040	3 uses	"	7.00	1.34	8.34
2080	5 uses	"	6.42	1.13	7.55
03110.20	**ELEVATED SLAB FORMWORK**				
0100	Elevated slab formwork				
1000	Slab, with drop panels				
1020	1 use	S.F.	5.04	4.63	9.67
1060	3 uses	"	4.66	2.08	6.74
1100	5 uses	"	4.34	1.66	6.00
2000	Floor slab, hung from steel beams				
2020	1 use	S.F.	4.84	3.73	8.57
2060	3 uses	"	4.50	1.86	6.36
2100	5 uses	"	4.20	1.37	5.57
3000	Floor slab, with pans or domes				
3020	1 use	S.F.	5.72	6.65	12.37
3060	3 uses	"	5.25	4.03	9.28
3100	5 uses	"	4.84	3.33	8.17
9030	Equipment curbs, 12" high				
9035	1 use	L.F.	6.30	3.43	9.73
9060	3 uses	"	5.72	1.92	7.64
9100	5 uses	"	5.25	1.49	6.74
03110.25	**EQUIPMENT PAD FORMWORK**				
1000	Equipment pad, job built				
1020	1 use	S.F.	7.87	4.50	12.37
1040	2 uses	"	7.41	2.70	10.11
1060	3 uses	"	7.00	2.16	9.16

		UNIT	LABOR	MAT.	TOTAL
03110.35	**FOOTING FORMWORK**				
2000	Wall footings, job built, continuous				
2040	1 use	S.F.	6.30	2.09	8.39
2060	3 uses	"	5.72	1.21	6.93
2090	5 uses	"	5.25	0.93	6.18
3000	Column footings, spread				
3020	1 use	S.F.	7.87	2.21	10.08
3060	3 uses	"	7.00	1.17	8.17
3100	5 uses	"	6.30	0.90	7.20
03110.50	**GRADE BEAM FORMWORK**				
1000	Grade beams, job built				
1020	1 use	S.F.	6.30	3.29	9.59
1060	3 uses	"	5.72	1.44	7.16
1100	5 uses	"	5.25	1.00	6.25
03110.53	**PILE CAP FORMWORK**				
1500	Pile cap forms, job built				
1510	Square				
1520	1 use	S.F.	7.87	3.74	11.61
1560	3 uses	"	7.00	1.71	8.71
1600	5 uses	"	6.30	1.25	7.55
03110.55	**SLAB / MAT FORMWORK**				
3000	Mat foundations, job built				
3020	1 use	S.F.	7.87	3.27	11.14
3060	3 uses	"	7.00	1.39	8.39
3100	5 uses	"	6.30	0.94	7.24
3980	Edge forms				
3990	6" high				
4000	1 use	L.F.	5.72	3.30	9.02
4002	3 uses	"	5.25	1.39	6.64
4004	5 uses	"	4.84	0.95	5.79
4014	5 uses	"	5.25	0.86	6.11
5000	Formwork for openings				
5020	1 use	S.F.	12.50	4.45	16.95
5060	3 uses	"	10.50	2.14	12.64
5100	5 uses	"	9.00	1.38	10.38
03110.60	**STAIR FORMWORK**				
1000	Stairway forms, job built				
1020	1 use	S.F.	12.50	5.14	17.64
1040	3 uses	"	10.50	2.24	12.74
1060	5 uses	"	9.00	1.72	10.72
03110.65	**WALL FORMWORK**				
2980	Wall forms, exterior, job built				
3000	Up to 8' high wall				
3120	1 use	S.F.	6.30	3.52	9.82
3160	3 uses	"	5.72	1.71	7.43
3190	5 uses	"	5.25	1.29	6.54
3200	Over 8' high wall				
3220	1 use	S.F.	7.87	3.87	11.74
3240	3 uses	"	7.00	2.01	9.01
3290	5 uses	"	6.30	1.58	7.88
5000	Column pier and pilaster				
5020	1 use	S.F.	12.50	3.87	16.37

		UNIT	LABOR	MAT.	TOTAL
03110.65	**WALL FORMWORK, Cont'd...**				
5060	3 uses	S.F.	10.50	2.13	12.63
5090	5 uses	"	9.00	1.74	10.74
6980	Interior wall forms				
7000	Up to 8' high				
7020	1 use	S.F.	5.72	3.52	9.24
7060	3 uses	"	5.25	1.74	6.99
7100	5 uses	"	4.84	1.25	6.09
7200	Over 8' high				
7220	1 use	S.F.	7.00	3.87	10.87
7260	3 uses	"	6.30	2.01	8.31
7290	5 uses	"	5.72	1.60	7.32
9000	PVC form liner, per side, smooth finish				
9010	1 use	S.F.	5.25	9.09	14.34
9030	3 uses	"	4.84	4.22	9.06
9050	5 uses	"	4.20	2.60	6.80
03110.70	**INSULATED CONCRETE FORMS**				
0010	4" thick, straight	S.F.	1.96	5.24	7.20
0020	90° corner	"	1.96	5.24	7.20
0030	45° angle	"	1.96	5.75	7.71
1000	6" thick, straight	"	1.96	5.31	7.27
1010	90° corner	"	1.96	5.31	7.27
1020	45° angle	"	1.96	5.75	7.71
1030	6" thick, corbel ledge	"	1.96	6.62	8.58
1040	T-block	"	1.96	5.94	7.90
2000	8" thick, straight	"	1.96	5.47	7.43
2010	90° corner	"	1.96	5.31	7.27
2020	45° angle	"	1.96	6.11	8.07
2030	corbel ledge	"	1.96	6.83	8.79
2040	T-block	"	1.96	6.28	8.24
03110.90	**MISCELLANEOUS FORMWORK**				
1200	Keyway forms (5 uses)				
1220	2 x 4	L.F.	3.15	0.30	3.45
1240	2 x 6	"	3.50	0.43	3.93
1500	Bulkheads				
1510	Walls, with keyways				
1515	2 piece	L.F.	5.72	4.97	10.69
1520	3 piece	"	6.30	6.28	12.58
1560	Elevated slab, with keyway				
1570	2 piece	L.F.	5.25	5.69	10.94
1580	3 piece	"	5.72	8.39	14.11
1600	Ground slab, with keyway				
1620	2 piece	L.F.	4.50	5.88	10.38
1640	3 piece	"	4.84	7.19	12.03
2000	Chamfer strips				
2020	Wood				
2040	1/2" wide	L.F.	1.40	0.28	1.68
2060	3/4" wide	"	1.40	0.36	1.76
2070	1" wide	"	1.40	0.49	1.89
2100	PVC				
2120	1/2" wide	L.F.	1.40	1.26	2.66
2140	3/4" wide	"	1.40	1.36	2.76

		UNIT	LABOR	MAT.	TOTAL
03110.90	**MISCELLANEOUS FORMWORK, Cont'd...**				
2160	1" wide	L.F.	1.40	1.98	3.38
2170	Radius				
2180	1"	L.F.	1.50	1.47	2.97
2200	1-1/2"	"	1.50	2.67	4.17
3000	Reglets				
3020	Galvanized steel, 24 ga.	L.F.	2.52	1.91	4.43
5000	Metal formwork				
5020	Straight edge forms				
5080	8" high	L.F.	4.50	35.25	39.75
5300	Curb form, S-shape				
5310	12" x				
5340	2'	L.F.	8.40	57.00	65.40
5380	3'	"	7.00	67.00	74.00
03210.05	**BEAM REINFORCING**				
0980	Beam-girders				
1000	#3 - #4	TON	1,610	1,630	3,240
1011	#7 - #8	"	1,070	1,360	2,430
1018	Galvanized				
1020	#3 - #4	TON	1,610	2,780	4,390
1031	#7 - #8	"	1,070	2,530	3,600
2000	Bond Beams				
2100	#3 - #4	TON	2,150	1,630	3,780
2120	#7 - #8	"	1,430	1,360	2,790
2200	Galvanized				
2210	#3 - #4	TON	2,150	2,660	4,810
2230	#7 - #8	"	1,430	2,530	3,960
03210.15	**COLUMN REINFORCING**				
0980	Columns				
1000	#3 - #4	TON	1,840	1,630	3,470
1015	#7 - #8	"	1,290	1,360	2,650
1100	Galvanized				
1200	#3 - #4	TON	1,840	2,780	4,620
1320	#7 - #8	"	1,290	2,530	3,820
03210.20	**ELEVATED SLAB REINFORCING**				
0980	Elevated slab				
1000	#3 - #4	TON	800	1,630	2,430
1040	#7 - #8	"	640	1,360	2,000
1980	Galvanized				
2000	#3 - #4	TON	800	2,660	3,460
2040	#7 - #8	"	640	2,530	3,170
03210.25	**EQUIP. PAD REINFORCING**				
0980	Equipment pad				
1000	#3 - #4	TON	1,290	1,630	2,920
1040	#7 - #8	"	1,070	1,360	2,430
03210.35	**FOOTING REINFORCING**				
1000	Footings				
1010	Grade 50				
1020	#3 - #4	TON	1,070	1,630	2,700
1040	#7 - #8	"	800	1,360	2,160
1055	Grade 60				
1060	#3 - #4	TON	1,070	1,630	2,700

DIVISION # 03 CONCRETE

		UNIT	LABOR	MAT.	TOTAL
03210.35	**FOOTING REINFORCING, Cont'd...**				
1074	#7 - #8	TON	800	1,360	2,160
4980	Straight dowels, 24" long				
5000	1" dia. (#8)	EA.	6.43	5.07	11.50
5040	3/4" dia. (#6)	"	6.43	4.55	10.98
5050	5/8" dia. (#5)	"	5.36	3.94	9.30
5060	1/2" dia. (#4)	"	4.59	2.97	7.56
03210.45	**FOUNDATION REINFORCING**				
0980	Foundations				
1000	#3 - #4	TON	1,070	1,630	2,700
1040	#7 - #8	"	800	1,360	2,160
1380	Galvanized				
1400	#3 - #4	TON	1,070	2,790	3,860
1420	#7 - #8	"	800	2,540	3,340
03210.50	**GRADE BEAM REINFORCING**				
0980	Grade beams				
1000	#3 - #4	TON	990	1,640	2,630
1040	#7 - #8	"	760	1,370	2,130
1090	Galvanized				
1100	#3 - #4	TON	990	2,790	3,780
1140	#7 - #8	"	760	2,540	3,300
03210.53	**PILE CAP REINFORCING**				
0980	Pile caps				
1000	#3 - #4	TON	1,610	1,640	3,250
1040	#7 - #8	"	1,290	1,370	2,660
1090	Galvanized				
1100	#3 - #4	TON	1,610	2,790	4,400
1140	#7 - #8	"	1,290	2,540	3,830
03210.55	**SLAB / MAT REINFORCING**				
0980	Bars, slabs				
1000	#3 - #4	TON	1,070	1,640	2,710
1040	#7 - #8	"	800	1,370	2,170
1980	Galvanized				
2000	#3 - #4	TON	1,070	2,790	3,860
2020	#5 - #6	"	920	2,640	3,560
2040	#7 - #8	"	800	2,540	3,340
5000	Wire mesh, slabs				
5010	Galvanized				
5015	4x4				
5020	W1.4xW1.4	S.F.	0.42	0.40	0.82
5040	W2.0xW2.0	"	0.45	0.52	0.97
5060	W2.9xW2.9	"	0.49	0.74	1.23
5080	W4.0xW4.0	"	0.53	1.09	1.62
5090	6x6				
5100	W1.4xW1.4	S.F.	0.32	0.37	0.69
5120	W2.0xW2.0	"	0.35	0.52	0.87
5140	W2.9xW2.9	"	0.37	0.71	1.08
5160	W4.0xW4.0	"	0.42	0.76	1.18
5170	Standard				
5175	2x2				
5180	W.9xW.9	S.F.	0.42	0.40	0.82
5185	4x4				

		UNIT	LABOR	MAT.	TOTAL
03210.55	**SLAB / MAT REINFORCING, Cont'd...**				
5190	W1.4xW1.4	S.F.	0.42	0.27	0.69
5500	W4.0xW4.0	"	0.53	0.74	1.27
5580	6x6				
5600	W1.4xW1.4	S.F.	0.32	0.17	0.49
6020	W4.0xW4.0	"	0.42	0.49	0.91
03210.60	**STAIR REINFORCING**				
0980	Stairs				
1000	#3 - #4	TON	1,290	1,640	2,930
1020	#5 - #6	"	1,070	1,440	2,510
1980	Galvanized				
2000	#3 - #4	TON	1,290	2,790	4,080
2020	#5 - #6	"	1,070	2,640	3,710
03210.65	**WALL REINFORCING**				
0980	Walls				
1000	#3 - #4	TON	920	1,640	2,560
1040	#7 - #8	"	720	1,370	2,090
1980	Galvanized				
2000	#3 - #4	TON	920	2,790	3,710
2040	#7 - #8	"	720	2,540	3,260
8980	Masonry wall (horizontal)				
9000	#3 - #4	TON	2,570	1,640	4,210
9020	#5 - #6	"	2,150	1,440	3,590
9030	Galvanized				
9040	#3 - #4	TON	2,570	2,790	5,360
9060	#5 - #6	"	2,150	2,630	4,780
9180	Masonry wall (vertical)				
9200	#3 - #4	TON	3,220	1,640	4,860
9220	#5 - #6	"	2,570	1,440	4,010
9230	Galvanized				
9240	#3 - #4	TON	3,220	2,790	6,010
9260	#5 - #6	"	2,570	2,630	5,200
03250.40	**CONCRETE ACCESSORIES**				
1000	Expansion joint, poured				
1010	Asphalt				
1020	1/2" x 1"	L.F.	0.98	0.87	1.85
1040	1" x 2"	"	1.07	2.71	3.78
1060	Liquid neoprene, cold applied				
1080	1/2" x 1"	L.F.	1.00	3.39	4.39
1100	1" x 2"	"	1.09	14.00	15.09
03300.10	**CONCRETE ADMIXTURES**				
1000	Concrete admixtures				
1020	Water reducing admixture	GAL			11.75
1040	Set retarder	"			25.25
1060	Air entraining agent	"			11.00
03350.10	**CONCRETE FINISHES**				
0980	Floor finishes				
1000	Broom	S.F.	0.70		0.70
1020	Screed	"	0.61		0.61
1040	Darby	"	0.61		0.61
1060	Steel float	"	0.82		0.82
4000	Wall finishes				

		UNIT	LABOR	MAT.	TOTAL
03350.10	**CONCRETE FINISHES, Cont'd...**				
4020	Burlap rub, with cement paste	S.F.	0.82	0.12	0.94
4160	Break ties and patch holes	"	0.98		0.98
4170	Carborundum				
4180	Dry rub	S.F.	1.64		1.64
4200	Wet rub	"	2.46		2.46
5000	Floor hardeners				
5010	Metallic				
5020	Light service	S.F.	0.61	0.42	1.03
5040	Heavy service	"	0.82	1.26	2.08
5050	Non-metallic				
5060	Light service	S.F.	0.61	0.21	0.82
5080	Heavy service	"	0.82	0.88	1.70
03360.10	**PNEUMATIC CONCRETE**				
0100	Pneumatic applied concrete (gunite)				
1035	2" thick	S.F.	3.40	6.18	9.58
1040	3" thick	"	4.53	7.59	12.12
1060	4" thick	"	5.44	9.25	14.69
1980	Finish surface				
2000	Minimum	S.F.	3.15		3.15
2020	Maximum	"	6.30		6.30
03370.10	**CURING CONCRETE**				
1000	Sprayed membrane				
1010	Slabs	S.F.	0.09	0.06	0.15
1020	Walls	"	0.12	0.08	0.20
1025	Curing paper				
1030	Slabs	S.F.	0.12	0.08	0.20
2000	Walls	"	0.14	0.08	0.22
2010	Burlap				
2020	7.5 oz.	S.F.	0.16	0.07	0.23
2500	12 oz.	"	0.17	0.09	0.26
03380.05	**BEAM CONCRETE**				
0960	Beams and girders				
0980	2500# or 3000# concrete				
1010	By pump	C.Y.	90.00	130	220
1020	By hand buggy	"	49.25	130	179
4000	5000# concrete				
4020	By pump	C.Y.	90.00	150	240
4040	By hand buggy	"	49.25	150	199
9460	Bond beam, 3000# concrete				
9470	By pump				
9480	8" high				
9500	4" wide	L.F.	1.98	0.36	2.34
9520	6" wide	"	2.25	0.87	3.12
9530	8" wide	"	2.47	1.10	3.57
9540	10" wide	"	2.75	1.47	4.22
9550	12" wide	"	3.09	1.98	5.07
03380.15	**COLUMN CONCRETE**				
0980	Columns				
0990	2500# or 3000# concrete				
1010	By pump	C.Y.	83.00	130	213
3980	5000# concrete				

		UNIT	LABOR	MAT.	TOTAL
03380.15	**COLUMN CONCRETE, Cont'd...**				
4020	By pump	C.Y.	83.00	150	233
03380.25	**EQUIPMENT PAD CONCRETE**				
0960	Equipment pad				
0980	2500# or 3000# concrete				
1000	By chute	C.Y.	16.50	130	147
1020	By pump	"	71.00	130	201
1050	3500# or 4000# concrete				
1060	By chute	C.Y.	16.50	130	147
1080	By pump	"	71.00	130	201
1110	5000# concrete				
1120	By chute	C.Y.	16.50	150	167
1140	By pump	"	71.00	150	221
03380.35	**FOOTING CONCRETE**				
0980	Continuous footing				
0990	2500# or 3000# concrete				
1000	By chute	C.Y.	16.50	130	147
1010	By pump	"	62.00	130	192
4000	5000# concrete				
4010	By chute	C.Y.	16.50	150	167
4020	By pump	"	62.00	150	212
4980	Spread footing				
5000	2500# or 3000# concrete				
5010	Under 5 cy				
5020	By chute	C.Y.	16.50	130	147
5040	By pump	"	66.00	130	196
7200	5000# concrete				
7205	Under 5 c.y.				
7210	By chute	C.Y.	16.50	140	157
7220	By pump	"	66.00	140	206
03380.50	**GRADE BEAM CONCRETE**				
0960	Grade beam				
0980	2500# or 3000# concrete				
1000	By chute	C.Y.	16.50	130	147
1040	By pump	"	62.00	130	192
1060	By hand buggy	"	49.25	130	179
1150	5000# concrete				
1160	By chute	C.Y.	16.50	150	167
1190	By pump	"	62.00	150	212
1200	By hand buggy	"	49.25	150	199
03380.53	**PILE CAP CONCRETE**				
0970	Pile cap				
0980	2500# or 3000 concrete				
1000	By chute	C.Y.	16.50	130	147
1010	By pump	"	71.00	130	201
1020	By hand buggy	"	49.25	130	179
3980	5000# concrete				
4010	By chute	C.Y.	16.50	150	167
4020	By pump	"	71.00	150	221
4030	By hand buggy	"	49.25	150	199

		UNIT	LABOR	MAT.	TOTAL
03380.55	**SLAB / MAT CONCRETE**				
0960	Slab on grade				
0980	2500# or 3000# concrete				
1000	By chute	C.Y.	12.25	130	142
1020	By pump	"	35.50	130	166
1030	By hand buggy	"	33.00	130	163
3980	5000# concrete				
4010	By chute	C.Y.	12.25	150	162
4030	By pump	"	35.50	150	186
4040	By hand buggy	"	33.00	150	183
03380.58	**SIDEWALKS**				
6000	Walks, cast in place with wire mesh, base not incl.				
6010	4" thick	S.F.	1.64	1.88	3.52
6020	5" thick	"	1.97	2.55	4.52
6030	6" thick	"	2.46	3.13	5.59
03380.60	**STAIR CONCRETE**				
0960	Stairs				
0980	2500# or 3000# concrete				
1000	By chute	C.Y.	16.50	130	147
1030	By pump	"	71.00	130	201
1040	By hand buggy	"	49.25	130	179
2100	3500# or 4000# concrete				
2120	By chute	C.Y.	16.50	140	157
2160	By pump	"	71.00	140	211
2180	By hand buggy	"	49.25	140	189
4000	5000# concrete				
4010	By chute	C.Y.	16.50	150	167
4030	By pump	"	71.00	150	221
4040	By hand buggy	"	49.25	150	199
03380.65	**WALL CONCRETE**				
0940	Walls				
0960	2500# or 3000# concrete				
0980	To 4'				
1000	By chute	C.Y.	14.00	130	144
1010	By pump	"	76.00	130	206
1020	To 8'				
1040	By pump	C.Y.	83.00	130	213
2960	3500# or 4000# concrete				
2980	To 4'				
3000	By chute	C.Y.	14.00	140	154
3030	By pump	"	76.00	140	216
3060	To 8'				
3100	By pump	C.Y.	83.00	140	223
8480	Filled block (CMU)				
8490	3000# concrete, by pump				
8500	4" wide	S.F.	3.54	0.50	4.04
8510	6" wide	"	4.13	1.12	5.25
8520	8" wide	"	4.95	1.76	6.71
8530	10" wide	"	5.83	2.35	8.18
8540	12" wide	"	7.08	3.02	10.10
8560	Pilasters, 3000# concrete	C.F.	99.00	6.87	106
8700	Wall cavity, 2" thick, 3000# concrete	S.F.	3.30	1.28	4.58

DIVISION # 03 CONCRETE

		UNIT	LABOR	MAT.	TOTAL
03550.10	**CONCRETE TOPPINGS**				
1000	Gypsum fill				
1020	2" thick	S.F.	0.51	1.97	2.48
1040	2-1/2" thick	"	0.52	2.25	2.77
1060	3" thick	"	0.53	2.77	3.30
1080	3-1/2" thick	"	0.55	3.16	3.71
1100	4" thick	"	0.61	3.70	4.31
2000	Formboard				
2020	Mineral fiber board				
2040	1" thick	S.F.	1.23	1.77	3.00
2060	1-1/2" thick	"	1.41	4.65	6.06
2070	Cement fiber board				
2080	1" thick	S.F.	1.64	1.38	3.02
2100	1-1/2" thick	"	1.89	1.77	3.66
2110	Glass fiber board				
2120	1" thick	S.F.	1.23	2.17	3.40
2140	1-1/2" thick	"	1.41	2.95	4.36
4000	Poured deck				
4010	Vermiculite or perlite				
4020	1 to 4 mix	C.Y.	83.00	190	273
4040	1 to 6 mix	"	76.00	170	246
4050	Vermiculite or perlite				
4060	2" thick				
4080	1 to 4 mix	S.F.	0.52	1.77	2.29
4100	1 to 6 mix	"	0.47	1.29	1.76
4200	3" thick				
4220	1 to 4 mix	S.F.	0.76	2.42	3.18
4240	1 to 6 mix	"	0.70	1.91	2.61
6000	Concrete plank, lightweight				
6020	2" thick	S.F.	4.07	9.61	13.68
6040	2-1/2" thick	"	4.07	9.87	13.94
6080	3-1/2" thick	"	4.52	10.25	14.77
6100	4" thick	"	4.52	10.75	15.27
6500	Channel slab, lightweight, straight				
6520	2-3/4" thick	S.F.	4.07	7.75	11.82
6540	3-1/2" thick	"	4.07	7.98	12.05
6560	3-3/4" thick	"	4.07	8.61	12.68
6580	4-3/4" thick	"	4.52	11.00	15.52
7000	Gypsum plank				
7020	2" thick	S.F.	4.07	3.58	7.65
7040	3" thick	"	4.07	3.76	7.83
8000	Cement fiber, T and G planks				
8020	1" thick	S.F.	3.70	1.97	5.67
8040	1-1/2" thick	"	3.70	2.10	5.80
8060	2" thick	"	4.07	2.50	6.57
8080	2-1/2" thick	"	4.07	2.66	6.73
8100	3" thick	"	4.07	3.45	7.52
8120	3-1/2" thick	"	4.52	3.99	8.51
8140	4" thick	"	4.52	4.38	8.90
03730.10	**CONCRETE REPAIR**				
0090	Epoxy grout floor patch, 1/4" thick	S.F.	4.93	8.38	13.31
0100	Grout, epoxy, 2 component system	C.F.			410

03730.10	CONCRETE REPAIR, Cont'd...	UNIT	LABOR	MAT.	TOTAL
0110	Epoxy sand	BAG			27.50
0120	Epoxy modifier	GAL			180
0140	Epoxy gel grout	S.F.	49.25	4.08	53.33
0150	Injection valve, 1 way, threaded plastic	EA.	9.87	11.25	21.12
0155	Grout crack seal, 2 component	C.F.	49.25	940	989
0160	Grout, non shrink	"	49.25	96.00	145
0165	Concrete, epoxy modified				
0170	Sand mix	C.F.	19.75	150	170
0180	Gravel mix	"	18.25	110	128
0190	Concrete repair				
0195	Soffit repair				
0200	16" wide	L.F.	9.87	4.78	14.65
0210	18" wide	"	10.25	5.09	15.34
0220	24" wide	"	11.00	6.08	17.08
0230	30" wide	"	11.75	6.84	18.59
0240	32" wide	"	12.25	7.29	19.54
0245	Edge repair				
0250	2" spall	L.F.	12.25	2.27	14.52
0260	3" spall	"	13.00	2.27	15.27
0270	4" spall	"	13.25	2.42	15.67
0280	6" spall	"	13.75	2.51	16.26
0290	8" spall	"	14.50	2.66	17.16
0300	9" spall	"	16.50	2.73	19.23
0330	Crack repair, 1/8" crack	"	4.93	4.48	9.41
5000	Reinforcing steel repair				
5005	1 bar, 4 ft				
5010	#4 bar	L.F.	8.04	0.70	8.74
5012	#5 bar	"	8.04	0.95	8.99
5014	#6 bar	"	8.58	1.15	9.73
5016	#8 bar	"	8.58	2.09	10.67
5020	#9 bar	"	9.19	2.67	11.86
5030	#11 bar	"	9.19	4.18	13.37
7010	Form fabric, nylon				
7020	18" diameter	L.F.			18.00
7030	20" diameter	"			18.50
7040	24" diameter	"			30.25
7050	30" diameter	"			31.00
7060	36" diameter	"			35.50
7100	Pile repairs				
7105	Polyethylene wrap				
7108	30 mil thick				
7110	60" wide	S.F.	16.50	19.50	36.00
7120	72" wide	"	19.75	21.50	41.25
7125	60 mil thick				
7130	60" wide	S.F.	16.50	23.25	39.75
7140	80" wide	"	22.50	26.75	49.25
8010	Pile spall, average repair 3'				
8020	18" x 18"	EA.	41.25	60.00	101
8030	20" x 20"	"	49.25	81.00	130

		UNIT	COST
03999.10	**CONCRETE**		
1000	FOOTINGS (Incl. exc. with steel)		
1010	By L.F - Continuous 24" x 12" (3 # 4 Rods)	L.F.	33.25
1020	36" x 12" (4 # 4 Rods)	"	42.25
1030	20" x 10" (2 # 5 Rods)	"	28.25
1040	16" x 8" (2 # 4 Rods)	"	24.25
1050	Pad 24" x 24" x 12" (4 # 5 E.W.)	EA.	115
1060	36" x 36" x 14" (6 # 5 EW.)	"	240
1070	48" x 48" x 16" (8 # 5 EW.)	"	420
1100	By C.Y. - Continuous 24" x 12" (3 # 4 Rods)	C.Y.	410
1110	36" x 12" (4 # 4 Rods)	"	440
1120	20" x 10" (2 # 5 Rods)	"	460
1130	16" x 8" (2 # 4 Rods)	"	560
1140	Pad 24" x 24" x 12" (4 # 5 E.W.)	"	690
1150	36" x 36" x 14" (6 # 5 E.W.)	"	650
1160	48" x 48" x 16" (8 # 5 E.W.)	"	620
1200	WALLS (#5 Rods 12" O.C. - 1 Face)		
1210	By S.F. - 8" Wall (# 5 12" O.C. E.W.)	S.F.	24.00
1220	12" Wall (# 5 12" O.C. E.W.)	"	25.50
1230	16" Wall (# 5 12" O.C. E.W.)	"	25.50
1300	Add for Steel - 2 Faces - 12" Wall	"	3.07
1310	Add for Pilastered Wall - 24" O.C.	"	1.46
1320	Add for Retaining or Battered Type	"	4.33
1330	Add for Curved Walls	"	8.03
1340	By C.Y. - 8" Wall (# 5 12" O.C. E.W.)	C.Y.	770
1350	12" Wall (# 5 12" O.C. E.W.)	"	640
1360	16" Wall (# 5 12" O.C. E.W.)	"	580
1400	Add for Steel - 2 Faces - 12" Wall	"	85.25
1410	Add for Pilastered Wall - 24" O.C.	"	37.50
1420	Add for Retaining or Battered Type	"	116
1430	Add for Curved Walls	"	211
1500	COLUMNS		
1510	By L.F. - Square Cornered 8" x 8" (4 # 8 Rods)	L.F.	54.50
1520	12" x 12" (6 # 8 Rods)	"	93.00
1530	16" x 16" (6 # 10 Rods)	"	120
1540	20" x 20" (8 # 20 Rods)	"	173
1550	24" x 24" (10 # 11 Rods)	"	211
1600	Round 8" (4 # 8 Rods)	"	40.00
1610	12" (6 # 8 Rods)	"	62.75
1620	16" (6 # 10 Rods)	"	85.50
1630	20" (8 # 20 Rods)	"	120
1640	24" (10 # 11 Rods)	"	174
1700	By C.Y. - Sq Cornered 8" x 8" (4 # 8 Rods)	C.Y.	3,170
1710	12" x 12" (6 # 8 Rods)	"	2,920
1720	16" x 16" (6 # 10 Rods)	"	2,080
1730	20" x 20" (8 # 20 Rods)	"	1,850
1740	24" x 24" (10 # 11 Rods)	"	1,320
1800	Round 8" (4 # 8 Rods)	"	2,920
1810	12" (6 # 8 Rods)	"	1,940
1820	16" (6 # 10 Rods)	"	1,790
1830	20" (8 # 20 Rods)	"	1,630
1840	24" (10 # 11 Rods)	"	1,550
1900	BEAMS		

		UNIT	COST
03999.10	**CONCRETE, Cont'd...**		
1910	By L.F. - Spandrel 12" x 48" (33 # rebar)	L.F.	211
1920	12" x 42" (26 # Rein. Steel)	"	191
1930	12" x 36" (21 # Rein. Steel)	"	151
1940	12" x 30" (15 # Rein. Steel)	"	143
1950	8" x 48" (26 # Rein. Steel)	"	187
1960	8" x 42" (21 # Rein. Steel)	"	170
1970	8" x 36" (16 # Rein. Steel)	"	153
2000	Interior 16" x 30" (24 # rebar)	"	121
2010	16" x 24" (20 # Rein. Steel)	"	120
2020	12" x 30" (17 # Rein. Steel)	"	141
2030	12" x 24" (14 # Rein. Steel)	"	101
2040	12" x 16" (12 # Rein. Steel)	"	85.75
2050	8" x 24" (13 # Rein. Steel)	"	101
2060	8" x 16" (10 # Rein. Steel)	"	75.75
2100	By C.Y. - Spandrel 12" x 48" (33 # rebar)	C.Y.	1,240
2110	12" x 42" (26 # Rein. Steel)	"	1,320
2120	12" x 36" (21 # Rein. Steel)	"	1,400
2130	12" x 30" (15 # Rein. Steel)	"	1,480
2140	8" x 48" (26 # Rein. Steel)	"	1,480
2150	8" x 42" (21 # Rein. Steel)	"	1,620
2160	8" x 36" (16 # Rein. Steel)	"	1,690
2200	Interior 16" x 30" (24 # rebar)	"	820
2210	16" x 24" (20 # Rein. Steel)	"	1,200
2220	12" x 30" (17 # Rein. Steel)	"	1,320
2230	12" x 24" (14 # Rein. Steel)	"	1,290
2240	12" x 16" (12 # Rein. Steel)	"	1,470
2250	8" x 24" (13 # Rein. Steel)	"	1,630
2260	8" x 16" (10 # Rein. Steel)	"	1,710
2300	SLABS (With Reinf. Steel)		
2310	By S.F. - Solid		
2320	4" Thick	S.F.	14.73
2330	5" Thick	"	17.25
2340	6" Thick	"	19.03
2350	7" Thick	"	21.22
2360	8" Thick	"	20.14
2400	Deduct for Post Tensioned Slabs	"	1.28
2410	Pan	"	
2420	Joist -20" Pan - 10" x 2"	"	19.39
2430	12" x 2"	"	22.00
2440	30" Pan - 10" x 2 1/2"	"	19.55
2450	12" x 2 1/2"	"	22.00
2500	Dome -19" x 19" - 10" x 2"	"	21.50
2510	12" x 2"	"	22.25
2520	30" x 30" - 10" x 2 1/2"	"	18.93
2530	12" x 2 1/2"	"	21.00
2600	By C.Y. - Solid		
2610	4" Thick	C.Y.	1,120
2620	5" Thick	"	1,070
2630	6" Thick	"	1,030
2640	8" Thick	"	1,000
2700	COMBINED COLUMNS, BEAMS AND SLABS		
2710	20' Span	S.F.	29.50

		UNIT	COST
03999.10	**CONCRETE, Cont'd...**		
2720	30' Span	S.F.	31.50
2730	40' Span	"	33.25
2740	50' Span	"	36.00
2750	60' Span	"	61.00
2800	STAIRS (Including Landing)		
2810	By EA. - 4' Wide 10' Floor Heights 16 Risers	EA.	270
2820	5' Wide 10' Floor Heights 16 Risers	"	300
2830	6' Wide 10' Floor Heights 16 Risers	"	360
2840	By C.Y. - 4' Wide 10' Floor Heights 16 Risers	C.Y.	1,630
2850	5' Wide 10' Floor Heights 16 Risers	"	1,680
2860	6' Wide 10' Floor Heights 16 Risers	"	1,650
2900	SLABS ON GROUND		
2910	4" Concrete Slab		
2920	(6 6/10 - 10 Mesh - 5 1/2 Sack Concrete,		
2930	Trowel Finished, Cured & Truck Chuted)	S.F.	5.01
2940	Add per Inch of Concrete	"	0.84
2950	Add per Sack of Cement	"	0.22
2960	Deduct for Float Finish	"	0.11
2970	Deduct for Brush or Broom Finish	"	0.09
3000	Add for Runway and Buggied Concrete	"	0.33
3010	Add for Vapor Barrier (4 mil)	"	0.23
3020	Add for Sub - Floor Fill (4" sand/gravel)	"	0.71
3030	Add for Change to 6 6/8 - 8 Mesh	"	0.12
3040	Add for Change to 6 6/6 - 6 Mesh	"	0.23
3050	Add for Sloped Slab	"	0.31
3060	Add for Edge Strip (sidewalk area)	"	0.33
3100	Add for ½" Expansion Joint (20' O.C.)	"	0.11
3110	Add for Control Joints (keyed/ dep.)	"	0.25
3120	Add for Control Joints (joint filled)	"	0.26
3130	Add for Control Joints - Saw Cut (20' O.C.)	"	0.47
3140	Add for Floor Hardener (1 coat)	"	0.22
3150	Add for Exposed Aggregate - Washed Added	"	0.63
3200	Retarding Added	"	0.60
3210	Seeding Added	"	0.62
3220	Add for Light Weight Aggregates	"	0.63
3230	Add for Heavy Weight Aggregates	"	0.43
3240	Add for Winter Production Loss/Cost	"	0.54
3300	TOPPING SLABS		
3310	2" Concrete	S.F.	4.27
3320	3" Concrete	"	5.32
3330	No Mesh or Hoisting		
3400	PADS & PLATFORMS (Including Form Work & Reinforcing)		
3410	4"	S.F.	10.44
3420	6"	"	13.19
3500	PRECAST CONCRETE ITEMS		
3510	Curbs 6" x 10" x 8"	L.F.	17.73
3520	Sills & Stools 6"	"	40.00
3530	Splash Blocks 3" x 16"	EA.	96.50
3600	MISCELLANEOUS ADDITIONS TO ABOVE CONCRETE		
3610	Abrasives - Carborundum - Grits	S.F.	1.38
3620	Strips	L.F.	4.51
3630	Bushhammer - Green Concrete	S.F.	3.04

		UNIT	COST
03999.10	**CONCRETE, Cont'd...**		
3640	Cured Concrete	S.F.	4.08
3650	Chamfers - Plastic 3/4"	L.F.	1.59
3660	Wood 3/4"	"	0.94
3670	Metal 3/4"	"	1.75
3680	Colors Dust On	S.F.	1.28
3690	Integral (Top 1")	"	2.15
3700	Control Joints - Asphalt 1/2" x 4"	L.F.	1.35
3710	1/2" x 6"	"	1.56
3720	PolyFoam 1/2" x 4"	"	1.69
3730	Dovetail Slots 22 Ga	"	2.36
3740	24 Ga	"	1.50
3800	Hardeners Acrylic and Urethane	S.F.	0.45
3810	Epoxy	"	0.57
3820	Joint Sealers Epoxy	L.F.	5.42
3830	Rubber Asphalt	"	2.87
3840	Moisture Proofing - Polyethylene 4 mil	S.F.	0.29
3850	6 mil	"	0.35
3900	Non-Shrink Grouts - Non - Metallic	C.F.	62.75
3910	Aluminum Oxide	"	45.50
3920	Iron Oxide	"	66.75
4000	Reglets Flashing	L.F.	4.18
4010	Sand Blast - Light	S.F.	2.15
4020	Heavy	"	4.19
4030	Shelf Angle Inserts 5/8"	L.F.	15.18
4040	Stair Nosings Steel - Galvanized	"	10.91
4050	Tongue & Groove Joint Forms - Asphalt 5 1/2"	"	3.27
4060	Wood 5 1/2"	"	2.68
4070	Metal 5 1/2"	"	3.41
4100	Treads - Extruded Aluminum	"	14.36
4110	Cast Iron	"	16.61
4120	Water Stops Center Bulb - Rubber - 6"	"	16.96
4130	9"	"	31.75
4140	Polyethylene - 6"	"	4.64
4150	9"	"	4.70

Design Cost Data™ DCD

TABLE OF CONTENTS | PAGE

		UNIT	LABOR	MAT.	TOTAL
04100.10	**MASONRY GROUT**				
0100	Grout, non shrink, non-metallic, trowelable	C.F.	1.81	5.09	6.90
2110	Grout door frame, hollow metal				1.10
2120	Single	EA.	68.00	12.75	80.75
2140	Double	"	72.00	17.75	89.75
2980	Grout-filled concrete block (CMU)				
3000	4" wide	S.F.	2.26	0.37	2.63
3020	6" wide	"	2.47	0.98	3.45
3040	8" wide	"	2.72	1.43	4.15
3060	12" wide	"	2.86	2.35	5.21
3070	Grout-filled individual CMU cells				
3090	4" wide	L.F.	1.36	0.31	1.67
3100	6" wide	"	1.36	0.42	1.78
3120	8" wide	"	1.36	0.55	1.91
3140	10" wide	"	1.55	0.70	2.25
3160	12" wide	"	1.55	0.84	2.39
4000	Bond beams or lintels, 8" deep				
4010	6" thick	L.F.	2.25	0.84	3.09
4020	8" thick	"	2.47	1.11	3.58
4040	10" thick	"	2.75	1.40	4.15
4060	12" thick	"	3.09	1.68	4.77
5000	Cavity walls				
5020	2" thick	S.F.	3.30	0.93	4.23
5040	3" thick	"	3.30	1.40	4.70
5060	4" thick	"	3.54	1.86	5.40
5080	6" thick	"	4.13	2.79	6.92
04150.10	**MASONRY ACCESSORIES**				
0200	Foundation vents	EA.	24.00	30.75	54.75
1010	Bar reinforcing				
1015	Horizontal				
1020	#3 - #4	Lb.	2.40	0.60	3.00
1030	#5 - #6	"	2.00	0.60	2.60
1035	Vertical				
1040	#3 - #4	Lb.	3.00	0.60	3.60
1050	#5 - #6	"	2.40	0.60	3.00
1100	Horizontal joint reinforcing				
1105	Truss type				
1110	4" wide, 6" wall	L.F.	0.24	0.20	0.44
1120	6" wide, 8" wall	"	0.25	0.20	0.45
1130	8" wide, 10" wall	"	0.26	0.25	0.51
1140	10" wide, 12" wall	"	0.27	0.25	0.52
1150	12" wide, 14" wall	"	0.28	0.30	0.58
1155	Ladder type				
1160	4" wide, 6" wall	L.F.	0.24	0.15	0.39
1170	6" wide, 8" wall	"	0.25	0.17	0.42
1180	8" wide, 10" wall	"	0.26	0.18	0.44
1190	10" wide, 12" wall	"	0.26	0.22	0.48
2000	Rectangular wall ties				
2005	3/16" dia., galvanized				
2010	2" x 6"	EA.	1.00	0.38	1.38
2020	2" x 8"	"	1.00	0.40	1.40
2040	2" x 10"	"	1.00	0.47	1.47

		UNIT	LABOR	MAT.	TOTAL
04150.10	**MASONRY ACCESSORIES, Cont'd...**				
2050	2" x 12"	EA.	1.00	0.53	1.53
2060	4" x 6"	"	1.20	0.44	1.64
2070	4" x 8"	"	1.20	0.49	1.69
2080	4" x 10"	"	1.20	0.63	1.83
2090	4" x 12"	"	1.20	0.73	1.93
2095	1/4" dia., galvanized				
2100	2" x 6"	EA.	1.00	0.71	1.71
2110	2" x 8"	"	1.00	0.80	1.80
2120	2" x 10"	"	1.00	0.91	1.91
2130	2" x 12"	"	1.00	1.04	2.04
2140	4" x 6"	"	1.20	0.82	2.02
2150	4" x 8"	"	1.20	0.91	2.11
2160	4" x 10"	"	1.20	1.04	2.24
2170	4" x 12"	"	1.20	1.08	2.28
2200	"Z" type wall ties, galvanized				
2215	6" long				
2220	1/8" dia.	EA.	1.00	0.34	1.34
2230	3/16" dia.	"	1.00	0.36	1.36
2240	1/4" dia.	"	1.00	0.38	1.38
2245	8" long				
2250	1/8" dia.	EA.	1.00	0.36	1.36
2260	3/16" dia.	"	1.00	0.38	1.38
2270	1/4" dia.	"	1.00	0.40	1.40
2275	10" long				
2280	1/8" dia.	EA.	1.00	0.38	1.38
2290	3/16" dia.	"	1.00	0.44	1.44
2300	1/4" dia.	"	1.00	0.49	1.49
3000	Dovetail anchor slots				
3015	Galvanized steel, filled				
3020	24 ga.	L.F.	1.50	1.02	2.52
3040	20 ga.	"	1.50	2.14	3.64
3060	16 oz. copper, foam filled	"	1.50	3.06	4.56
3100	Dovetail anchors				
3115	16 ga.				
3120	3-1/2" long	EA.	1.00	0.34	1.34
3140	5-1/2" long	"	1.00	0.41	1.41
3150	12 ga.				
3160	3-1/2" long	EA.	1.00	0.45	1.45
3180	5-1/2" long	"	1.00	0.74	1.74
3200	Dovetail, triangular galvanized ties, 12 ga.				
3220	3" x 3"	EA.	1.00	0.77	1.77
3240	5" x 5"	"	1.00	0.82	1.82
3260	7" x 7"	"	1.00	0.93	1.93
3280	7" x 9"	"	1.00	0.99	1.99
3400	Brick anchors				
3420	Corrugated, 3-1/2" long				
3440	16 ga.	EA.	1.00	0.33	1.33
3460	12 ga.	"	1.00	0.59	1.59
3500	Non-corrugated, 3-1/2" long				
3520	16 ga.	EA.	1.00	0.47	1.47
3540	12 ga.	"	1.00	0.85	1.85
3580	Cavity wall anchors, corrugated, galvanized				

		UNIT	LABOR	MAT.	TOTAL
04150.10	**MASONRY ACCESSORIES, Cont'd...**				
3600	5" long				
3620	16 ga.	EA.	1.00	0.95	1.95
3640	12 ga.	"	1.00	1.43	2.43
3660	7" long				
3680	28 ga.	EA.	1.00	1.05	2.05
3700	24 ga.	"	1.00	1.33	2.33
3720	22 ga.	"	1.00	1.36	2.36
3740	16 ga.	"	1.00	1.54	2.54
3800	Mesh ties, 16 ga., 3" wide				
3820	8" long	EA.	1.00	1.28	2.28
3840	12" long	"	1.00	1.43	2.43
3860	20" long	"	1.00	1.96	2.96
3900	24" long	"	1.00	2.16	3.16
04150.20	**MASONRY CONTROL JOINTS**				
1000	Control joint, cross shaped PVC	L.F.	1.50	2.38	3.88
1010	Closed cell joint filler				
1020	1/2"	L.F.	1.50	0.41	1.91
1040	3/4"	"	1.50	0.85	2.35
1070	Rubber, for				
1080	4" wall	L.F.	1.50	2.75	4.25
1090	6" wall	"	1.58	3.40	4.98
1100	8" wall	"	1.67	4.10	5.77
1110	PVC, for				
1120	4" wall	L.F.	1.50	1.43	2.93
1140	6" wall	"	1.58	2.41	3.99
1160	8" wall	"	1.67	3.65	5.32
04150.50	**MASONRY FLASHING**				
0080	Through-wall flashing				
1000	5 oz. coated copper	S.F.	5.01	3.88	8.89
1020	0.030" elastomeric	"	4.00	1.27	5.27
04210.10	**BRICK MASONRY**				
0100	Standard size brick, running bond				
1000	Face brick, red (6.4/sf)				
1020	Veneer	S.F.	10.00	5.66	15.66
1030	Cavity wall	"	8.58	5.66	14.24
1040	9" solid wall	"	17.25	11.25	28.50
1200	Common brick (6.4/sf)				
1210	Select common for veneers	S.F.	10.00	3.69	13.69
1215	Back-up				
1220	4" thick	S.F.	7.51	3.32	10.83
1230	8" thick	"	12.00	6.65	18.65
1235	Firewall				
1240	12" thick	S.F.	20.00	10.75	30.75
1250	16" thick	"	27.25	14.25	41.50
1300	Glazed brick (7.4/sf)				
1310	Veneer	S.F.	11.00	15.25	26.25
1400	Buff or gray face brick (6.4/sf)				
1410	Veneer	S.F.	10.00	6.58	16.58
1420	Cavity wall	"	8.58	6.58	15.16
1500	Jumbo or oversize brick (3/sf)				
1510	4" veneer	S.F.	6.01	4.33	10.34

		UNIT	LABOR	MAT.	TOTAL
04210.10	**BRICK MASONRY, Cont'd...**				
1530	4" back-up	S.F.	5.01	4.33	9.34
1540	8" back-up	"	8.58	5.02	13.60
1550	12" firewall	"	15.00	6.75	21.75
1560	16" firewall	"	20.00	9.52	29.52
1600	Norman brick, red face, (4.5/sf)				
1620	4" veneer	S.F.	7.51	7.17	14.68
1640	Cavity wall	"	6.68	7.17	13.85
3000	Chimney, standard brick, including flue				
3020	16" x 16"	L.F.	60.00	32.00	92.00
3040	16" x 20"	"	60.00	54.00	114
3060	16" x 24"	"	60.00	58.00	118
3080	20" x 20"	"	75.00	45.00	120
3100	20" x 24"	"	75.00	61.00	136
3120	20" x 32"	"	86.00	68.00	154
4000	Window sill, face brick on edge	"	15.00	3.59	18.59
04210.20	**STRUCTURAL TILE**				
5000	Structural glazed tile				
5010	6T series, 5-1/2" x 12"				
5020	Glazed on one side				
5040	2" thick	S.F.	6.01	11.00	17.01
5060	4" thick	"	6.01	13.25	19.26
5080	6" thick	"	6.68	20.75	27.43
5100	8" thick	"	7.51	25.50	33.01
5200	Glazed on two sides				
5220	4" thick	S.F.	7.51	19.25	26.76
5240	6" thick	"	8.58	26.50	35.08
5500	Special shapes				
5510	Group 1	S.F.	12.00	11.25	23.25
5520	Group 2	"	12.00	14.25	26.25
5530	Group 3	"	12.00	18.75	30.75
5540	Group 4	"	12.00	37.75	49.75
5550	Group 5	"	12.00	46.00	58.00
5600	Fire rated				
5620	4" thick, 1 hr rating	S.F.	6.01	18.00	24.01
5640	6" thick, 2 hr rating	"	6.68	25.25	31.93
6000	8W series, 8" x 16"				
6010	Glazed on one side				
6020	2" thick	S.F.	4.00	12.50	16.50
6040	4" thick	"	4.00	13.25	17.25
6060	6" thick	"	4.62	22.00	26.62
6080	8" thick	"	4.62	24.00	28.62
6100	Glazed on two sides				
6120	4" thick	S.F.	5.01	21.00	26.01
6140	6" thick	"	6.01	29.25	35.26
6160	8" thick	"	6.01	35.50	41.51
6200	Special shapes				
6220	Group 1	S.F.	8.58	18.75	27.33
6230	Group 2	"	8.58	23.00	31.58
6240	Group 3	"	8.58	25.25	33.83
6250	Group 4	"	8.58	42.00	50.58
6260	Group 5	"	8.58	53.00	61.58

		UNIT	LABOR	MAT.	TOTAL
04210.20	**STRUCTURAL TILE, Cont'd...**				
6270	Fire rated				
6290	4" thick, 1 hr rating	S.F.	8.58	31.50	40.08
6300	6" thick, 2 hr rating	"	8.58	43.25	51.83
04210.60	**PAVERS, MASONRY**				
4000	Brick walk laid on sand, sand joints				
4020	Laid flat, (4.5 per sf)	S.F.	6.68	4.16	10.84
4040	Laid on edge, (7.2 per sf)	"	10.00	6.66	16.66
5000	Precast concrete patio blocks				
5005	2" thick				
5010	Natural	S.F.	2.00	3.58	5.58
5020	Colors	"	2.00	4.53	6.53
5080	Exposed aggregates, local aggregate				
5100	Natural	S.F.	2.00	8.91	10.91
5120	Colors	"	2.00	8.91	10.91
5130	Granite or limestone aggregate	"	2.00	9.31	11.31
5140	White tumblestone aggregate	"	2.00	6.66	8.66
6000	Stone pavers, set in mortar				
6005	Bluestone				
6008	1" thick				
6010	Irregular	S.F.	15.00	8.57	23.57
6020	Snapped rectangular	"	12.00	13.00	25.00
6060	1-1/2" thick, random rectangular	"	15.00	15.00	30.00
6070	2" thick, random rectangular	"	17.25	17.75	35.00
6090	Slate				
6100	Natural cleft				
6110	Irregular, 3/4" thick	S.F.	17.25	10.75	28.00
6115	Random rectangular				
6120	1-1/4" thick	S.F.	15.00	23.25	38.25
6130	1-1/2" thick	"	16.75	26.25	43.00
7000	Granite blocks				
7010	3" thick, 3" to 6" wide				
7020	4" to 12" long	S.F.	20.00	13.25	33.25
7030	6" to 15" long	"	17.25	8.67	25.92
9000	Flagstone pavers				
9010	Random sizes, 1-4 s.f., tumbled cobble	S.F.	12.00	17.00	29.00
9020	Tumbled patio	"	12.00	18.25	30.25
9040	Saw-cut Flagstone Tiles				
9060	12"x12", assorted colors	S.F.	8.58	15.75	24.33
9080	18"x18", assorted colors	"	7.51	15.75	23.26
9100	24"x24", assorted colors	"	6.68	15.75	22.43
9800	Crushed stone, white marble, 3" thick	"	0.98	1.88	2.86
04220.10	**CONCRETE MASONRY UNITS**				
0110	Hollow, load bearing				
0120	4"	S.F.	4.45	1.55	6.00
0140	6"	"	4.62	2.27	6.89
0160	8"	"	5.01	2.60	7.61
0180	10"	"	5.46	3.59	9.05
0190	12"	"	6.01	4.13	10.14
0280	Solid, load bearing				
0300	4"	S.F.	4.45	2.43	6.88
0320	6"	"	4.62	2.72	7.34

		UNIT	LABOR	MAT.	TOTAL
04220.10	**CONCRETE MASONRY UNITS, Cont'd...**				
0340	8"	S.F.	5.01	3.72	8.73
0360	10"	"	5.46	3.97	9.43
0380	12"	"	6.01	5.89	11.90
0480	Back-up block, 8" x 16"				
0500	2"	S.F.	3.43	1.62	5.05
0540	4"	"	3.53	1.69	5.22
0560	6"	"	3.75	2.48	6.23
0580	8"	"	4.00	2.84	6.84
0600	10"	"	4.29	3.93	8.22
0620	12"	"	4.62	4.52	9.14
0980	Foundation wall, 8" x 16"				
1000	6"	S.F.	4.29	2.48	6.77
1030	8"	"	4.62	2.84	7.46
1040	10"	"	5.01	3.93	8.94
1050	12"	"	5.46	4.53	9.99
1055	Solid				
1060	6"	S.F.	4.62	3.00	7.62
1070	8"	"	5.01	4.09	9.10
1080	10"	"	5.46	4.35	9.81
1100	12"	"	6.01	6.47	12.48
1480	Exterior, styrofoam inserts, std weight, 8" x 16"				
1500	6"	S.F.	4.62	4.35	8.97
1530	8"	"	5.01	4.70	9.71
1540	10"	"	5.46	6.09	11.55
1550	12"	"	6.01	8.36	14.37
1580	Lightweight				
1600	6"	S.F.	4.62	4.85	9.47
1660	8"	"	5.01	5.45	10.46
1680	10"	"	5.46	5.80	11.26
1700	12"	"	6.01	7.65	13.66
1980	Acoustical slotted block				
2000	4"	S.F.	5.46	5.06	10.52
2020	6"	"	5.46	5.30	10.76
2040	8"	"	6.01	6.61	12.62
2050	Filled cavities				
2060	4"	S.F.	6.68	5.42	12.10
2070	6"	"	7.07	6.24	13.31
2080	8"	"	7.51	8.00	15.51
4000	Hollow, split face				
4020	4"	S.F.	4.45	3.46	7.91
4030	6"	"	4.62	4.01	8.63
4040	8"	"	5.01	4.21	9.22
4080	10"	"	5.46	4.71	10.17
4100	12"	"	6.01	5.02	11.03
4480	Split rib profile				
4500	4"	S.F.	5.46	4.21	9.67
4520	6"	"	5.46	4.89	10.35
4540	8"	"	6.01	5.32	11.33
4560	10"	"	6.01	5.83	11.84
4580	12"	"	6.01	6.31	12.32
4980	High strength block, 3500 psi				
5000	2"	S.F.	4.45	1.63	6.08

		UNIT	LABOR	MAT.	TOTAL
04220.10	**CONCRETE MASONRY UNITS, Cont'd...**				
5020	4"	S.F.	4.62	2.04	6.66
5030	6"	"	4.62	2.44	7.06
5040	8"	"	5.01	2.76	7.77
5050	10"	"	5.46	3.22	8.68
5060	12"	"	6.01	3.81	9.82
5500	Solar screen concrete block				
5505	4" thick				
5510	6" x 6"	S.F.	13.25	4.29	17.54
5520	8" x 8"	"	12.00	5.12	17.12
5530	12" x 12"	"	9.24	5.24	14.48
5540	8" thick				
5550	8" x 16"	S.F.	8.58	5.24	13.82
7000	Glazed block				
7020	Cove base, glazed 1 side, 2"	L.F.	6.68	10.50	17.18
7030	4"	"	6.68	10.75	17.43
7040	6"	"	7.51	11.00	18.51
7050	8"	"	7.51	11.75	19.26
7055	Single face				
7060	2"	S.F.	5.01	11.00	16.01
7080	4"	"	5.01	13.50	18.51
7090	6"	"	5.46	14.50	19.96
7100	8"	"	6.01	15.25	21.26
7105	10"	"	6.68	17.25	23.93
7110	12"	"	7.07	18.25	25.32
7115	Double face				
7120	4"	S.F.	6.32	16.25	22.57
7140	6"	"	6.68	19.25	25.93
7160	8"	"	7.51	20.25	27.76
7180	Corner or bullnose				
7200	2"	EA.	7.51	17.50	25.01
7240	4"	"	8.58	22.25	30.83
7260	6"	"	8.58	27.25	35.83
7280	8"	"	10.00	29.75	39.75
7290	10"	"	11.00	32.25	43.25
7300	12"	"	12.00	34.75	46.75
9500	Gypsum unit masonry				
9510	Partition blocks (12"x30")				
9515	Solid				
9520	2"	S.F.	2.40	1.41	3.81
9525	Hollow				
9530	3"	S.F.	2.40	1.42	3.82
9540	4"	"	2.50	1.63	4.13
9550	6"	"	2.73	1.74	4.47
9900	Vertical reinforcing				
9920	4' o.c., add 5% to labor				
9940	2'8" o.c., add 15% to labor				
9960	Interior partitions, add 10% to labor				
04220.90	**BOND BEAMS & LINTELS**				
0980	Bond beam, no grout or reinforcement				
0990	8" x 16" x				
1000	4" thick	L.F.	4.62	1.85	6.47

		UNIT	LABOR	MAT.	TOTAL
04220.90	**BOND BEAMS & LINTELS, Cont'd...**				
1040	6" thick	L.F.	4.80	2.83	7.63
1060	8" thick	"	5.01	3.24	8.25
1080	10" thick	"	5.22	4.01	9.23
1100	12" thick	"	5.46	4.56	10.02
6000	Beam lintel, no grout or reinforcement				
6010	8" x 16" x				
6020	10" thick	L.F.	6.01	8.85	14.86
6040	12" thick	"	6.68	9.42	16.10
6080	Precast masonry lintel				
7000	6 lf, 8" high x				
7020	4" thick	L.F.	10.00	7.80	17.80
7040	6" thick	"	10.00	9.96	19.96
7060	8" thick	"	11.00	11.25	22.25
7080	10" thick	"	11.00	13.50	24.50
7090	10 lf, 8" high x				
7100	4" thick	L.F.	6.01	9.80	15.81
7120	6" thick	"	6.01	12.00	18.01
7140	8" thick	"	6.68	13.50	20.18
7160	10" thick	"	6.68	18.25	24.93
8000	Steel angles and plates				
8010	Minimum	Lb.	0.85	1.21	2.06
8020	Maximum	"	1.50	1.77	3.27
8200	Various size angle lintels				
8205	1/4" stock				
8210	3" x 3"	L.F.	3.75	6.23	9.98
8220	3" x 3-1/2"	"	3.75	6.86	10.61
8225	3/8" stock				
8230	3" x 4"	L.F.	3.75	10.75	14.50
8240	3-1/2" x 4"	"	3.75	11.25	15.00
8250	4" x 4"	"	3.75	12.50	16.25
8260	5" x 3-1/2"	"	3.75	13.25	17.00
8262	6" x 3-1/2"	"	3.75	14.75	18.50
8265	1/2" stock				
8280	6" x 4"	L.F.	3.75	16.50	20.25
04240.10	**CLAY TILE**				
0100	Hollow clay tile, for back-up, 12" x 12"				
1000	Scored face				
1010	Load bearing				
1020	4" thick	S.F.	4.29	6.56	10.85
1040	6" thick	"	4.45	7.65	12.10
1060	8" thick	"	4.62	9.53	14.15
1080	10" thick	"	4.80	11.75	16.55
1100	12" thick	"	5.01	20.00	25.01
2000	Non-load bearing				
2020	3" thick	S.F.	4.14	5.32	9.46
2040	4" thick	"	4.29	6.16	10.45
2060	6" thick	"	4.45	7.12	11.57
2080	8" thick	"	4.62	9.08	13.70
2100	12" thick	"	5.01	16.00	21.01
4100	Partition, 12" x 12"				
4150	In walls				

		UNIT	LABOR	MAT.	TOTAL
04240.10	**CLAY TILE, Cont'd...**				
4201	3" thick	S.F.	5.01	5.32	10.33
4210	4" thick	"	5.01	6.17	11.18
4220	6" thick	"	5.22	6.79	12.01
4230	8" thick	"	5.46	8.92	14.38
4240	10" thick	"	5.72	10.50	16.22
4250	12" thick	"	6.01	15.50	21.51
4300	Clay tile floors				
4320	4" thick	S.F.	3.34	6.16	9.50
4330	6" thick	"	3.53	7.65	11.18
4340	8" thick	"	3.75	9.53	13.28
4350	10" thick	"	4.00	11.75	15.75
4360	12" thick	"	4.29	17.50	21.79
6000	Terra cotta				
6020	Coping, 10" or 12" wide, 3" thick	L.F.	12.00	14.75	26.75
04270.10	**GLASS BLOCK**				
1000	Glass block, 4" thick				
1040	6" x 6"	S.F.	20.00	38.25	58.25
1060	8" x 8"	"	15.00	24.25	39.25
1080	12" x 12"	"	12.00	30.75	42.75
8980	Replacement glass blocks, 4" x 8" x 8"				
9100	Minimum	S.F.	60.00	17.75	77.75
9120	Maximum	"	120	30.50	151
04295.10	**PARGING / MASONRY PLASTER**				
0080	Parging				
0100	1/2" thick	S.F.	4.00	0.33	4.33
0200	3/4" thick	"	5.01	0.40	5.41
0300	1" thick	"	6.01	0.56	6.57
04400.10	**STONE**				
0160	Rubble stone				
0180	Walls set in mortar				
0200	8" thick	S.F.	15.00	16.50	31.50
0220	12" thick	"	24.00	20.00	44.00
0420	18" thick	"	30.00	26.50	56.50
0440	24" thick	"	40.00	33.25	73.25
0445	Dry set wall				
0450	8" thick	S.F.	10.00	18.75	28.75
0455	12" thick	"	15.00	21.00	36.00
0460	18" thick	"	20.00	29.00	49.00
0465	24" thick	"	24.00	35.50	59.50
0480	Cut stone				
0490	Imported marble				
0510	Facing panels				
0520	3/4" thick	S.F.	24.00	45.25	69.25
0530	1-1/2" thick	"	27.25	64.00	91.25
0540	2-1/4" thick	"	33.50	77.00	111
0600	Base				
0610	1" thick				
0620	4" high	L.F.	30.00	20.50	50.50
0640	6" high	"	30.00	24.75	54.75
0700	Columns, solid				
0720	Plain faced	C.F.	400	140	540

		UNIT	LABOR	MAT.	TOTAL
04400.10	**STONE, Cont'd...**				
0740	Fluted	C.F.	400	400	800
0780	Flooring, travertine, minimum	S.F.	9.24	20.00	29.24
0800	Average	"	12.00	26.50	38.50
0820	Maximum	"	13.25	48.50	61.75
1000	Domestic marble				
1020	Facing panels				
1040	7/8" thick	S.F.	24.00	40.50	64.50
1060	1-1/2" thick	"	27.25	61.00	88.25
1080	2-1/4" thick	"	33.50	74.00	108
1500	Stairs				
1510	12" treads	L.F.	30.00	36.25	66.25
1520	6" risers	"	20.00	26.75	46.75
1525	Thresholds, 7/8" thick, 3' long, 4" to 6" wide				
1530	Plain	EA.	50.00	33.00	83.00
1540	Beveled	"	50.00	36.25	86.25
1545	Window sill				
1550	6" wide, 2" thick	L.F.	24.00	18.25	42.25
1555	Stools				
1560	5" wide, 7/8" thick	L.F.	24.00	24.50	48.50
1620	Limestone panels up to 12' x 5', smooth finish				
1630	2" thick	S.F.	11.00	29.50	40.50
1650	3" thick	"	11.00	34.50	45.50
1660	4" thick	"	11.00	49.25	60.25
1760	Miscellaneous limestone items				
1770	Steps, 14" wide, 6" deep	L.F.	40.00	63.00	103
1780	Coping, smooth finish	C.F.	20.00	88.00	108
1790	Sills, lintels, jambs, smooth finish	"	24.00	88.00	112
1800	Granite veneer facing panels, polished				
1810	7/8" thick				
1820	Black	S.F.	24.00	48.25	72.25
1840	Gray	"	24.00	38.00	62.00
1850	Base				
1860	4" high	L.F.	12.00	20.25	32.25
1870	6" high	"	13.25	24.50	37.75
1880	Curbing, straight, 6" x 16"	"	45.50	22.50	68.00
1890	Radius curbs, radius over 5'	"	61.00	27.50	88.50
1900	Ashlar veneer				
1905	4" thick, random	S.F.	24.00	34.00	58.00
1910	Pavers, 4" x 4" split				
1915	Gray	S.F.	12.00	33.25	45.25
1920	Pink	"	12.00	32.75	44.75
1930	Black	"	12.00	32.25	44.25
2000	Slate, panels				
2010	1" thick	S.F.	24.00	25.00	49.00
2020	2" thick	"	27.25	34.00	61.25
2030	Sills or stools				
2040	1" thick				
2060	6" wide	L.F.	24.00	11.75	35.75
2080	10" wide	"	26.25	19.00	45.25
2100	2" thick				
2120	6" wide	L.F.	27.25	19.00	46.25
2140	10" wide	"	30.00	31.50	61.50

		UNIT	COST
04999.10	**MASONRY**		
1000	BRICK MASONRY		
1100	Conventional (Modular Size)		
1110	Running Bond - 8" x 2 2/3" x 4"	S.F.	20.41
1120	Common Bond - 6 Course Header	"	23.68
1130	Stack Bond	"	20.76
1140	Dutch & English Bond - Every Other Course Header	"	30.25
1150	Every Course Header	"	33.75
1160	Flemish Bond - Every Other Course Header	"	22.91
1170	Every Course Header	"	27.25
1200	Add: If Scaffold Needed	"	1.02
1210	Add: For each 10' of Floor Hgt. or Floor (3%)	"	0.57
1220	Add: Piers and Corbels (15% to Labor)	"	2.91
1230	Add: Sills and Soldiers (20% to Labor)	"	2.78
1240	Add: Floor Brick (10% to Labor)	"	2.03
1300	Add: Weave & Herringbone Patterns (20% to Labor)	"	2.77
1310	Add: Stack Bond (8% to Labor)	"	1.60
1320	Add: Circular or Radius Work 20% to Labor)	"	4.24
1330	Add: Rock Faced & Slurried Face (10% to Labor)	"	2.03
1340	Add: Arches (75% to Labor)	"	6.24
1350	Add: For Winter Work (below 40°)	"	2.03
1400	Production Loss (10% to Labor)	"	2.03
1410	Enclosures - Wall Area Conventional	"	2.03
1420	Heat and Fuel - Wall Area Conventional	"	0.41
1500	Econo - 8" x 4" x 3"	"	16.80
1600	Panel - 8" x 8" x 4"	"	13.17
1700	Norman - 12" x 2 2/3" x 4"	"	16.83
1800	King Size - 10" x 2 5/8" x 4"	"	14.60
1900	Norwegian - 12" x 3 1/5" x 4"	"	14.07
2000	Saxon-Utility - 12" x 4" x 3"	"	13.20
2010	12" x 4" x 4"	"	13.24
2020	12" x 4" x 6"	"	16.76
2030	12" x 4" x 8"	"	20.07
2200	Adobe - 12" x 3" x 4"	"	22.76
2300	COATED BRICK (Ceramic)	"	27.25
2400	COMMON BRICK (Clay, Concrete and Sand Lime)	"	15.54
2500	FIRE BRICK -Light Duty	"	25.51
2510	Heavy Duty	"	32.50
2520	Deduct for Residential Work - All Above - 5%		
04999.20	**CONCRETE BLOCK**		
1000	CONVENTIONAL (Struck 2 Sides - Partitions)		
1010	12" x 8"x 16" Plain	S.F.	10.55
1020	Bond Beam (with Fill-Reinforcing)	"	13.96
1030	12" x 8"x 8" Half Block	"	12.69
1040	Double End - Header	"	10.81
1050	8" x 8" x 16" Plain	"	9.21
1060	Bond Beam (with Fill-Reinforcing)	"	11.30
1070	8" x 8" x 8" Half Block	"	9.02
1080	Double End - Header	"	9.58
1090	6 "x 8" x 16" Plain	"	8.59
2000	Bond Beam (with Fill-Reinforcing)	"	9.63
2010	Half Block	"	8.47

		UNIT	COST
04999.20	**CONCRETE BLOCK, Cont'd...**		
2020	4" x 8" x 16" Plain	S.F.	8.31
2030	Half Block	"	7.24
2040	16" x 8" x 16" Plain	"	11.96
2050	Half Block	"	11.40
2060	14" x 8" x 16" Plain	"	11.35
2070	Half Block	"	10.78
2080	10" x 8" x 16" Plain	"	9.84
3000	Bond Beam (with Fill-Reinforcing)	"	11.33
3010	Half Block	"	11.03
3020	Deduct: Block Struck or Cleaned One Side	"	0.34
3030	Deduct: Block Not Struck or Cleaned Two Sides	"	0.55
3040	Deduct: Clean One Side Only	"	0.25
3050	Deduct: Lt. Wt. Block (Labor Only)	"	0.22
4000	Add: Jamb and Sash Block	"	0.58
4010	Add: Bullnose Block	"	0.86
4020	Add: Pilaster, Pier, Pedestal Work	"	0.95
4030	Add: Stack Bond Work	"	0.35
4040	Add: Radius or Circular Work	"	1.78
5000	Add: If Scaffold Needed	"	0.97
5010	Add: Winter Production Cost (10% Labor)	"	0.57
5020	Add: Winter Enclosing - Wall Area	"	0.57
5030	Add: Winter Heating - Wall Area	"	0.52
5040	Add: Core & Beam Filling - See 0411 (Page 4A-18)		
5050	Add: Insulation - See 0412 (Page 4A-18)		
6000	Deduct for Residential Work - 5%		
8010	SCREEN WALL - 4" x 12" x 12"	S.F.	8.28
8020	BURNISHED - 12" x 8" x 16"	"	15.51
8030	8" x 8" x 16"	"	14.45
8040	6" x 8" x 16"	"	13.20
8050	4" x 8" x 16"	"	12.39
8060	2" x 8" x 16"	"	11.16
8070	Add for Shapes	"	3.36
8080	Add for 2 Faced Finish	"	4.84
8090	Add for Scored Finish - 1 Face	"	1.17
8100	PREFACED UNITS -12" x 8" x 16" Stretcher	"	21.67
8110	(Ceramic Glazed) 12" x 8" x 16" Glazed 2 Face	"	29.25
8120	8" x 8" x 16" Stretcher	"	20.34
8130	Glazed 2 Face	"	27.50
8140	6" x 8" x 16" Stretcher	"	18.03
8150	Glazed 2 Face	"	27.50
8160	4" x 8" x 16" Stretcher	"	18.91
8170	Glazed 2 Face	"	26.40
8180	2" x 8" x 16" Stretcher	"	18.77
8190	4" x 16" x 16" Stretcher	"	48.50
8200	Add for Base, Caps, Jambs, Headers, Lintels	"	5.44
8210	Add for Scored Block	"	1.60
04999.30	**CLAY BACKING AND PARTITION TILE**		
0010	3" x 12" x 12"	S.F.	7.57
0020	4" x 12" x 12"	"	8.11
0030	6" x 12" x 12"	"	9.67
0040	8" x 12" x 12"	"	10.59

		UNIT	COST
04999.40	**CLAY FACING TILE (GLAZED)**		
1000	6T or 5" x 12" - SERIES		
1010	2" x 5 1/3" x 12" Soap Stretcher (Solid Back)	S.F.	27.25
1020	4" x 5 1/2" x 12" 1 Face Stretcher	"	29.00
1030	2 Face Stretcher	"	34.25
1040	6" x 5 1/3" x 12" 1 Face Stretcher	"	35.00
1050	8" x 5 1/3" x 12" 1 Face Stretcher	"	19.85
2000	8W or 8" x 16" - SERIES		
2010	2" x 8" x 16" Soap Stretcher (Solid Back)	S.F.	20.83
2020	4" x 8" x 16" 1 Face Stretcher	"	21.34
2030	2 Face Stretcher	"	28.00
2040	6" x 8" x 16" 1 Face Stretcher	"	25.15
2050	8" x 8" x 16" 1 Face Stretcher	"	26.25
2060	Add for Shapes (Average)	PCT.	57.75
2070	Add for Designer Colors	"	23.10
2080	Add for Less than Truckload Lots	"	11.55
2090	Add for Base Only	"	28.88
04999.50	**GLASS UNITS**		
0010	4" x 8" x 4"	S.F.	69.00
0020	6" x 6" x 4"	"	73.50
0030	8" x 8" x 4"	"	44.50
0040	12" x 12" x 4"	"	45.00
04999.60	**TERRA-COTTA**		
0010	Unglazed	S.F.	15.45
0020	Glazed	"	20.90
0030	Colored Glazed	"	25.10
04999.70	**FLUE LINING**		
0010	8" x 12"	S.F.	25.75
0020	12" x 12"	"	27.75
0030	16" x 16"	"	44.50
0040	18" x 18"	"	45.25
0050	20" x 20"	"	89.25
0060	24" x 24"	"	114
04999.80	**NATURAL STONE**		
1000	CUT STONE BY S.F.		
1010	Limestone- Indiana and Alabama - 3"	S.F.	52.25
1020	4"	"	57.75
1030	Minnesota, Wisc, Texas, etc.- 3"	"	58.50
1040	4"	"	67.00
1200	Marble - 2"	"	72.75
1210	3"	"	77.00
1300	Granite - 2"	"	67.75
1310	3"	"	73.00
1400	Slate - 1 1/2"	"	72.50
2000	Ashlar - 4" Sawed Bed		
2100	Limestone- Indiana - Random	S.F.	41.00
2110	Coursed - 2" - 5" - 8"	"	39.50
2120	Minnesota, Alabama, Wisconsin, etc.		
2130	Split Face - Random	S.F.	41.00
2140	Coursed	"	46.25
2150	Sawed or Planed Face - Random	"	44.25
2160	Coursed	"	46.75

		UNIT	COST
04999.80	**NATURAL STONE, Cont'd...**		
2200	Marble - Sawed - Random	S.F.	58.00
2210	Coursed	"	59.00
2300	Granite - Bushhammered - Random	"	59.25
2310	Coursed	"	66.00
2400	Quartzite	"	45.50
3000	ROUGH STONE		
3100	Rubble and Flagstone	S.F.	42.00
3200	Field Stone or Boulders	"	39.50
3300	Light Weight Boulders (Igneous)		
3310	2" to 4" Veneer - Sawed Back	S.F.	33.75
3320	3" to 10" Boulders	"	36.00
3330	CUT STONE BY C.F		
3400	Limestone- Indiana and Alabama - 3"	C.F.	233
3410	4"	"	166
3420	Minnesota, Wisc, Texas, etc.- 3"	"	240
3430	4"	"	230
3500	Marble - 2"	"	500
3510	3"	"	310
3600	Granite - 2"	"	460
3610	3"	"	310
3700	Slate - 1 1/2"	"	470
3800	ASHLAR Limestone- Indiana - Random	"	121
3810	Coursed - 2" - 5" - 8"	"	121
3820	Split Face - Random	"	123
3830	Coursed	"	151
3840	Sawed or Planed Face - Random	"	150
3850	Coursed	"	153
3900	Marble - Sawed - Random	"	187
3910	Coursed	"	189
4000	Granite - Bushhammered - Random	"	190
4010	Coursed	"	195
4100	Quartzite	"	150
4200	ROUGH STONE, Rubble and Flagstone	"	123
4300	Field Stone or Boulders	"	121
04999.90	**PRECAST VENEERS AND SIMULATED MASONRY**		
1000	ARCHITECTURAL PRECAST STONE		
1005	Limestone	S.F.	48.50
1010	Marble	"	58.00
2010	PRECAST CONCRETE	"	39.50
3010	MOSAIC GRANITE PANELS	"	54.75
04999.91	**MORTAR**		
0010	Portland Cement and Lime Mortar - Labor in Unit Costs	S.F.	148
0020	Masonry Cement Mortar - Labor in Unit Costs	"	142
0030	Mortar for Standard Brick - Material Only	"	0.61
04999.92	**CORE FILLING FOR REINFORCED CONCRETE**		
1010	Add to Block Prices in 0402.0 (Job Mixed)		
1020	12" x 8" x 16" Plain (includes #4 Rod 16" O.C. Vertical)	S.F.	4.70
1030	Bond Beam (includes 2 #4 Rods)	"	5.54
1040	8" x 8" x 6" Plain (includes #4 Rod 16" O.C. Vertical)	"	2.87
1050	Bond Beam (includes 2 #4 Rods)	"	2.33
1060	6" x 8" x 16" Plain	"	2.54

		UNIT	COST
04999.92	**CORE FILLING FOR REINFORCED CONCRETE BLOCK, Cont'd...**		
1070	Bond Beam (includes 1 #4 Rod)	S.F.	1.30
1080	10" x 8" x 16" Bond Beam (includes 2 #4 Rods)	"	5.03
1090	14" x 8" x 16" Bond Beam (includes 2 #4 Rods)	"	5.74
04999.93	**CORE AND CAVITY FILL FOR INSULATED**		
2100	LOOSE		
2110	Core Fill - Expanded Styrene		
2120	12" x 8" x 16" Concrete Block	S.F.	1.74
2130	10" x 8" x 16" Concrete Block	"	1.55
2140	8" x 8" x 16" Concrete Block	"	1.11
2150	6" x 8" x 16" Concrete Block	"	0.95
2160	8" Brick - Jumbo - Thru the Wall	"	0.84
2170	6" Brick - Jumbo - Thru the Wall	"	0.68
2180	4" Brick - Jumbo	"	0.58
2190	Cavity Fill - per Inch		
2200	Expanded Styrene	S.F.	0.74
2210	Mica	"	1.24
2220	Fiber Glass	"	0.77
2230	Rock Wool	"	0.77
2240	Cellulose	"	0.74
2300	RIGID - Fiber Glass - 3# Density, 1"	"	1.29
2310	2"1.58 -		
2320	Expanded Styrene - Molded - 1# Density, 1"	S.F.	0.95
2330	2"	"	1.41
2340	Extruded - 2# Density, 1"	"	1.36
2350	2"	"	2.07
2360	Expanded Urethane - 1"	"	1.46
2370	2"	"	1.94
2380	Perlite - 1"	"	1.41
2390	2"	"	2.20
2400	Add for Embedded Water Type	"	2.20
2410	Add for Glued Applications	"	0.08
04999.95	**CLEANING AND POINTING**		
4010	Brick and Stone - Acid or Chemicals	S.F.	1.02
4020	Soap and Water	"	0.97
4030	Block & Facing Tile - 1 Face (Incl. Hollow Metal Frames)	"	0.52
4040	Point with White Cement	"	2.04
04999.97	**SCAFFOLD**		
7100	EQUIPMENT, TOOLS AND BLADES		
7110	Percentage of Labor as an average	PCT.	7.35
7120	See Division 1 for Rental Rates & New Costs		
7200	SCAFFOLD- Tubular Frame - to 40 feet - Exterior	S.F.	1.28
7210	Tubular Frame - to 16 feet - Interior	"	1.22
7220	Swing Stage - 40 feet and up	"	1.36

TABLE OF CONTENTS   PAGE

		UNIT	LABOR	MAT.	TOTAL
05050.10	**STRUCTURAL WELDING**				
0080	Welding				
0100	Single pass				
0120	1/8"	L.F.	3.48	0.33	3.81
0140	3/16"	"	4.64	0.55	5.19
0160	1/4"	"	5.80	0.77	6.57
05050.30	**MECHANICAL ANCHORS**				
0010	Drop-in, stainless steel, for masonry, 3/8"	EA.			1.40
0020	1/2"	"			1.55
0030	Hollow wall, for use in gypsum drywall, 6-32"x				
0040	1-1/4"	EA.			0.33
0050	1-1/2"	"			0.38
0060	1-3/4"	"			0.35
0070	2"	"			0.33
0080	2-3/8"	"			0.35
1000	Concrete anchor, 3/16"x				
1010	1-1/4"	EA.			0.44
1020	1-3/4"	"			0.55
1030	2-3/4"	"			0.55
1040	1/4"x				
1050	1-1/4"	EA.			0.55
1060	2-1/4"	"			0.66
1070	2-3/4"	"			0.71
1080	Toggle anchor, 5/8"	"			1.50
2000	3/4"	"			1.50
2010	Toggle bolts, 1/8"x				
2020	2"	EA.			0.27
2030	3"	"			0.27
2040	4"	"			0.27
2050	3/16"x				
2060	2"	EA.			0.55
2070	3"	"			0.55
2080	4"	"			0.55
2090	1/4"x				
3000	2"	EA.			0.80
3010	3"	"			0.80
3020	4"	"			0.80
3030	6"	"			1.61
3050	Sleeve anchor, 3/8"x				
3060	1-7/8"	EA.			0.45
3070	2-1/4"	"			0.50
3080	3"	"			0.58
3090	1/4x2-1/4"	"			0.52
4000	1/2"x				
4010	2-1/4"	EA.			0.58
4020	3"	"			0.66
4030	4"	"			0.88
4040	5/8"x				
4050	2-1/4"	EA.			1.02
4060	4-1/4"	"			1.02
4070	6"	"			1.10
4080	Machine screw, corrosion resistant, for use in masonry, 5/8"x				

		UNIT	LABOR	MAT.	TOTAL
05050.30	**MECHANICAL ANCHORS, Cont'd...**				
4090	2"	EA.			1.21
5000	4"	"			1.82
5010	6"	"			3.08
5020	8"	"			4.47
5030	Wedge anchor, 1/4"x2-1/4"	"			0.50
5040	3/8"x				
5050	2-1/4"	EA.			0.50
5060	3"	"			0.66
5070	4"	"			0.72
5080	5"	"			0.88
5090	1/2"x				
6000	2-1/2"	EA.			0.70
6010	3-1/4"	"			1.28
6020	4-1/4"	"			1.52
6030	5-1/2"	"			1.90
6040	5/8"x				
6050	6"	EA.			2.72
6060	7"	"			3.20
6070	8"	"			3.68
6080	Spring wing, toggle bolt, for use in hollow walls, 1/8"x				
6090	2"	EA.			0.39
7000	3"	"			0.39
7010	4"	"			0.39
7020	3/16"x				
7030	2"	EA.			0.55
7040	3"	"			0.55
7050	4"	"			0.55
7060	5"	"			0.55
7070	1/4"x				
7080	2"	EA.			0.80
7090	3"	"			0.80
8000	4"	"			0.80
8010	3/8"x6"	"			1.61
8020	Stud anchor, 1-1/4"	"			0.55
8030	3/4"	"			0.55
8040	Hex bolt, zinc plated, 3/4"x				
8050	4"	EA.			2.91
8060	6"	"			3.57
8070	8"	"			4.66
8080	1/2"x				
8090	4"	EA.			0.94
9000	6"	"			1.46
9010	8"	"			1.85
9020	Lag screw, 1/4"x				
9030	2"	EA.			0.22
9040	4"	"			0.49
9050	6"	"			0.77
9060	Machine screw, zinc plated, 5/8"x				
9070	4"	EA.			1.45
9100	6"	"			2.45
9110	8"	"			3.55
9120	Ribbed plastic anchor 5/8"x				

		UNIT	LABOR	MAT.	TOTAL
05050.30	**MECHANICAL ANCHORS, Cont'd...**				
9130	1-1/4"	EA.			0.28
9140	1-1/2"	"			0.41
05050.90	**METAL ANCHORS**				
1000	Anchor bolts, material only				
1020	3/8" x				
1040	8" long	EA.			1.01
1080	12" long	"			1.19
1090	1/2" x				
1100	8" long	EA.			1.50
1140	12" long	"			1.76
1170	5/8" x				
1180	8" long	EA.			1.40
1220	12" long	"			1.65
1270	3/4" x				
1280	8" long	EA.			2.00
1300	12" long	"			2.25
4480	Non-drilling anchor				
4500	1/4"	EA.			0.64
4540	3/8"	"			0.80
4560	1/2"	"			1.23
7000	Self-drilling anchor				
7020	1/4"	EA.			1.62
7060	3/8"	"			2.44
7080	1/2"	"			3.25
05050.95	**METAL LINTELS**				
0080	Lintels, steel				
0100	Plain	Lb.	1.74	1.33	3.07
0120	Galvanized	"	1.74	2.00	3.74
05120.10	**BEAMS, GIRDERS, COLUMNS, TRUSSES**				
0100	Beams and girders, A-36				
0120	Welded	TON	810	2,870	3,680
0140	Bolted	"	740	2,800	3,540
0180	Columns				
0185	Pipe				
0190	6" dia.	Lb.	0.81	1.61	2.42
1300	Structural tube				
1310	6" square				
1320	Light sections	TON	1,630	3,770	5,400
05410.10	**METAL FRAMING**				
0100	Furring channel, galvanized				
0110	Beams and columns, 3/4"				
0120	12" o.c.	S.F.	6.96	0.44	7.40
0140	16" o.c.	"	6.32	0.34	6.66
0150	Walls, 3/4"				
0160	12" o.c.	S.F.	3.48	0.44	3.92
0170	16" o.c.	"	2.90	0.34	3.24
0172	24" o.c.	"	2.32	0.24	2.56
0173	1-1/2"				
0174	12" o.c.	S.F.	3.48	0.72	4.20
0175	16" o.c.	"	2.90	0.55	3.45
0176	24" o.c.	"	2.32	0.37	2.69

05410.10	METAL FRAMING, Cont'd...	UNIT	LABOR	MAT.	TOTAL
0177	Stud, load bearing				
0178	16" o.c.				
0179	16 ga.				
0180	2-1/2"	S.F.	3.09	1.33	4.42
0190	3-5/8"	"	3.09	1.57	4.66
0200	4"	"	3.09	1.63	4.72
0220	6"	"	3.48	2.05	5.53
0280	18 ga.				
0300	2-1/2"	S.F.	3.09	1.08	4.17
0310	3-5/8"	"	3.09	1.33	4.42
0320	4"	"	3.09	1.39	4.48
0330	6"	"	3.48	1.76	5.24
0350	8"	"	3.48	2.12	5.60
0360	20 ga.				
0370	2-1/2"	S.F.	3.09	0.60	3.69
0390	3-5/8"	"	3.09	0.72	3.81
0400	4"	"	3.09	0.79	3.88
0420	6"	"	3.48	0.96	4.44
0440	8"	"	3.48	1.15	4.63
0480	24" o.c.				
0490	16 ga.				
0500	2-1/2"	S.F.	2.67	0.91	3.58
0510	3-5/8"	"	2.67	1.08	3.75
0520	4"	"	2.67	1.15	3.82
0530	6"	"	2.90	1.39	4.29
0540	8"	"	2.90	1.76	4.66
0545	18 ga.				
0550	2-1/2"	S.F.	2.67	0.72	3.39
0560	3-5/8"	"	2.67	0.84	3.51
0570	4"	"	2.67	0.91	3.58
0580	6"	"	2.90	1.15	4.05
0590	8"	"	2.90	1.39	4.29
0595	20 ga.				
0600	2-1/2"	S.F.	2.67	0.44	3.11
0610	3-5/8"	"	2.67	0.49	3.16
0620	4"	"	2.67	0.55	3.22
0630	6"	"	2.90	0.71	3.61
0640	8"	"	2.90	0.88	3.78
05510.10	**STAIRS**				
1000	Stock unit, steel, complete, per riser				
1010	Tread				
1020	3'-6" wide	EA.	87.00	210	297
1040	4' wide	"	99.00	240	339
1060	5' wide	"	120	280	400
1200	Metal pan stair, cement filled, per riser				
1220	3'-6" wide	EA.	70.00	230	300
1240	4' wide	"	77.00	260	337
1260	5' wide	"	87.00	290	377
1280	Landing, steel pan	S.F.	17.50	90.00	108
1300	Cast iron tread, steel stringers, stock units, per riser				
1310	Tread				

		UNIT	LABOR	MAT.	TOTAL
05510.10	**STAIRS, Cont'd...**				
1320	3'-6" wide	EA.	87.00	400	487
1340	4' wide	"	99.00	460	559
1360	5' wide	"	120	550	670
1400	Stair treads, abrasive, 12" x 3'-6"				
1410	Cast iron				
1420	3/8"	EA.	34.75	190	225
1440	1/2"	"	34.75	240	275
1450	Cast aluminum				
1460	5/16"	EA.	34.75	230	265
1480	3/8"	"	34.75	240	275
1500	1/2"	"	34.75	290	325
05515.10	**LADDERS**				
0100	Ladder, 18" wide				
0110	With cage	L.F.	46.50	110	157
0120	Without cage	"	34.75	70.00	105
05520.10	**RAILINGS**				
0080	Railing, pipe				
0090	1-1/4" diameter, welded steel				
0095	2-rail				
0100	Primed	L.F.	14.00	31.75	45.75
0120	Galvanized	"	14.00	40.75	54.75
0130	3-rail				
0140	Primed	L.F.	17.50	40.75	58.25
0160	Galvanized	"	17.50	53.00	70.50
0170	Wall mounted, single rail, welded steel				
0180	Primed	L.F.	10.75	21.25	32.00
0200	Galvanized	"	10.75	27.50	38.25
0210	1-1/2" diameter, welded steel				
0215	2-rail				
0220	Primed	L.F.	14.00	34.50	48.50
0240	Galvanized	"	14.00	44.75	58.75
0245	3-rail				
0250	Primed	L.F.	17.50	43.25	60.75
0260	Galvanized	"	17.50	56.00	73.50
0270	Wall mounted, single rail, welded steel				
0280	Primed	L.F.	10.75	21.75	32.50
0300	Galvanized	"	10.75	28.50	39.25
0960	2" diameter, welded steel				
0980	2-rail				
1000	Primed	L.F.	15.50	41.25	56.75
1020	Galvanized	"	15.50	54.00	69.50
1030	3-rail				
1040	Primed	L.F.	20.00	52.00	72.00
1070	Galvanized	"	20.00	68.00	88.00
1075	Wall mounted, single rail, welded steel				
1080	Primed	L.F.	11.50	23.75	35.25
1100	Galvanized	"	11.50	30.75	42.25
05580.10	**METAL SPECIALTIES**				
0060	Kick plate				
0080	4" high x 1/4" thick				
0100	Primed	L.F.	14.00	8.47	22.47

		UNIT	LABOR	MAT.	TOTAL
05580.10	**METAL SPECIALTIES, Cont'd...**				
0120	Galvanized	L.F.	14.00	9.61	23.61
0130	6" high x 1/4" thick				
0140	Primed	L.F.	15.50	8.71	24.21
0160	Galvanized	"	15.50	10.25	25.75
0200	Fire Escape (10'-12' high)				
0210	Landing with fixed stair, 3'-6" wide	EA.	1,390	5,580	6,970
0220	4'-6" wide	"	1,390	6,310	7,700
05700.10	**ORNAMENTAL METAL**				
1020	Railings, square bars, 6" o.c., shaped top rails				
1040	Steel	L.F.	34.75	92.00	127
1060	Aluminum	"	34.75	110	145
1080	Bronze	"	46.50	190	237
1100	Stainless steel	"	46.50	190	237
1200	Laminated metal or wood handrails				
1220	2-1/2" round or oval shape	L.F.	34.75	280	315

		UNIT	COST
05999.10	**STRUCTURAL STEEL FRAME**		
1000	To 30 Ton 20' Span	S.F.	12.21
1010	24' Span	"	12.88
1020	28' Span	"	13.02
1030	32' Span	"	15.56
1040	36' Span	"	17.28
2010	44' Span	"	19.88
2020	48' Span	"	21.30
2030	52' Span	"	23.69
2050	60' Span	"	27.66
3000	Deduct for Over 30 Ton	PCT.	0.05
05999.40	**OPEN WEB JOISTS**		
1000	To 20 Ton, 20' Span	S.F.	4.60
1010	24' Span	"	4.63
1030	32' Span	"	4.70
2000	40' Span	"	5.35
2010	48' Span	"	6.28
2030	56' Span	"	7.28
2040	60' Span	"	7.80
3000	Deduct for Over 20 Ton	PCT.	0.10
05999.50	**METAL DECKING**		
1000	1/2" Deep - Ribbed - Baked Enamel - 18 Ga	S.F.	5.76
1020	22 Ga	"	4.85
2000	3" Deep - Ribbed - Baked Enamel - 18 Ga	"	11.47
2020	22 Ga	"	10.10
3000	4 1/2" Deep - Ribbed - Baked Enamel - 16 Ga	"	15.07
3010	18 Ga	"	13.11
3020	20 Ga	"	12.14
4000	3" Deep - Cellular - 18 Ga	"	13.74
4010	16 Ga	"	16.89
4020	4 1/2" Deep - Cellular - 18 Ga	"	19.28
4030	16 Ga	"	22.23
4040	Add for Galvanized	PCT	16.50
4050	Corrugated Black Standard .015	S.F.	2.65
4060	Heavy Duty - 26 Ga	"	2.86
4070	S. Duty - 24 Ga	"	3.75
4080	22 Ga	"	3.89
5000	Add for Galvanized	PCT.	15.00
05999.60	**METAL SIDING AND ROOFING**		
1000	Aluminum- Anodized	S.F.	6.99
1010	Porcelainized	"	13.90
1020	Corrugated	"	4.74
1030	Enamel, Baked Ribbed	"	7.41
1040	24 Ga Ribbed	"	5.08
1050	Acrylic	"	8.67
2000	Porcelain Ribbed	"	9.56
2010	Galvanized- Corrugated	"	4.77
2020	Plastic Faced	"	10.40
2030	Protected Metal	"	12.67
2040	Add for Liner Panels	"	3.93
2050	Add for Insulation	"	0.60

TABLE OF CONTENTS   PAGE

		UNIT	LABOR	MAT.	TOTAL
06110.10	**BLOCKING**				
1215	Wood construction				
1220	Walls				
1230	2x4	L.F.	3.50	0.60	4.10
1240	2x6	"	3.93	0.92	4.85
1250	2x8	"	4.20	1.21	5.41
1260	2x10	"	4.50	1.62	6.12
1270	2x12	"	4.84	2.09	6.93
1280	Ceilings				
1290	2x4	L.F.	3.93	0.60	4.53
1300	2x6	"	4.50	0.92	5.42
1310	2x8	"	4.84	1.21	6.05
1320	2x10	"	5.25	1.62	6.87
1330	2x12	"	5.72	2.09	7.81
06110.20	**CEILING FRAMING**				
1000	Ceiling joists				
1070	16" o.c.				
1080	2x4	S.F.	1.21	0.73	1.94
1090	2x6	"	1.26	1.09	2.35
1100	2x8	"	1.31	1.55	2.86
1110	2x10	"	1.36	1.74	3.10
1120	2x12	"	1.43	3.26	4.69
1130	24" o.c.				
1140	2x4	S.F.	1.00	0.52	1.52
1150	2x6	"	1.05	0.87	1.92
1160	2x8	"	1.10	1.30	2.40
1170	2x10	"	1.16	1.55	2.71
1180	2x12	"	1.23	3.92	5.15
1200	Headers and nailers				
1210	2x4	L.F.	2.03	0.60	2.63
1220	2x6	"	2.10	0.92	3.02
1230	2x8	"	2.25	1.21	3.46
1240	2x10	"	2.42	1.62	4.04
1250	2x12	"	2.62	1.99	4.61
1300	Sister joists for ceilings				
1310	2x4	L.F.	4.50	0.60	5.10
1320	2x6	"	5.25	0.92	6.17
1330	2x8	"	6.30	1.21	7.51
1340	2x10	"	7.87	1.62	9.49
1350	2x12	"	10.50	1.99	12.49
06110.30	**FLOOR FRAMING**				
1000	Floor joists				
1180	16" o.c.				
1190	2x6	S.F.	1.05	0.95	2.00
1200	2x8	"	1.06	1.34	2.40
1220	2x10	"	1.08	1.63	2.71
1230	2x12	"	1.12	2.03	3.15
1240	2x14	"	1.16	4.52	5.68
1250	3x6	"	1.08	3.13	4.21
1260	3x8	"	1.12	4.00	5.12
1270	3x10	"	1.16	5.04	6.20
1280	3x12	"	1.21	6.09	7.30

		UNIT	LABOR	MAT.	TOTAL
06110.30	**FLOOR FRAMING, Cont'd...**				
1290	3x14	S.F.	1.26	7.21	8.47
1300	4x6	"	1.08	4.00	5.08
1310	4x8	"	1.12	5.48	6.60
1320	4x10	"	1.16	6.78	7.94
1330	4x12	"	1.21	8.00	9.21
1340	4x14	"	1.26	9.57	10.83
2000	Sister joists for floors				
2010	2x4	L.F.	3.93	0.60	4.53
2020	2x6	"	4.50	0.92	5.42
2030	2x8	"	5.25	1.21	6.46
2040	2x10	"	6.30	1.62	7.92
2050	2x12	"	7.87	2.09	9.96
2060	3x6	"	6.30	3.04	9.34
2070	3x8	"	7.00	3.73	10.73
2080	3x10	"	7.87	4.95	12.82
2090	3x12	"	9.00	5.99	14.99
2100	4x6	"	6.30	3.92	10.22
2110	4x8	"	7.00	5.22	12.22
2120	4x10	"	7.87	6.78	14.65
2130	4x12	"	9.00	7.56	16.56
06110.40	**FURRING**				
1100	Furring, wood strips				
1102	Walls				
1105	On masonry or concrete walls				
1107	1x2 furring				
1110	12" o.c.	S.F.	1.96	0.48	2.44
1120	16" o.c.	"	1.80	0.41	2.21
1130	24" o.c.	"	1.65	0.40	2.05
1135	1x3 furring				
1140	12" o.c.	S.F.	1.96	0.60	2.56
1150	16" o.c.	"	1.80	0.55	2.35
1160	24" o.c.	"	1.65	0.42	2.07
1165	On wood walls				
1167	1x2 furring				
1170	12" o.c.	S.F.	1.40	0.48	1.88
1180	16" o.c.	"	1.26	0.41	1.67
1190	24" o.c.	"	1.14	0.38	1.52
1195	1x3 furring				
1200	12" o.c.	S.F.	1.40	0.62	2.02
1210	16" o.c.	"	1.26	0.52	1.78
1220	24" o.c.	"	1.14	0.42	1.56
06110.50	**ROOF FRAMING**				
1000	Roof framing				
1005	Rafters, gable end				
1008	0-2 pitch (flat to 2-in-12)				
1070	16" o.c.				
1080	2x6	S.F.	1.12	1.09	2.21
1090	2x8	"	1.16	1.53	2.69
1100	2x10	"	1.21	1.74	2.95
1110	2x12	"	1.26	3.21	4.47
1120	24" o.c.				

		UNIT	LABOR	MAT.	TOTAL
06110.50	**ROOF FRAMING, Cont'd...**				
1130	2x6	S.F.	0.95	0.60	1.55
1140	2x8	"	0.98	1.27	2.25
1150	2x10	"	1.01	1.48	2.49
1160	2x12	"	1.05	2.60	3.65
1165	4-6 pitch (4-in-12 to 6-in-12)				
1220	16" o.c.				
1230	2x6	S.F.	1.16	1.09	2.25
1240	2x8	"	1.21	1.74	2.95
1250	2x10	"	1.26	1.99	3.25
1260	2x12	"	1.31	2.96	4.27
1270	24" o.c.				
1280	2x6	S.F.	0.98	0.87	1.85
1290	2x8	"	1.01	1.48	2.49
1300	2x10	"	1.08	1.56	2.64
1310	2x12	"	1.21	2.43	3.64
1315	8-12 pitch (8-in-12 to 12-in-12)				
1380	16" o.c.				
1390	2x6	S.F.	1.21	1.21	2.42
1400	2x8	"	1.26	1.95	3.21
1410	2x10	"	1.31	2.17	3.48
1420	2x12	"	1.36	3.13	4.49
1430	24" o.c.				
1440	2x6	S.F.	1.01	0.95	1.96
1450	2x8	"	1.05	1.55	2.60
1460	2x10	"	1.08	1.74	2.82
1470	2x12	"	1.12	2.78	3.90
2000	Ridge boards				
2010	2x6	L.F.	3.15	0.92	4.07
2020	2x8	"	3.50	1.21	4.71
2030	2x10	"	3.93	1.62	5.55
2040	2x12	"	4.50	2.09	6.59
3000	Hip rafters				
3010	2x6	L.F.	2.25	0.92	3.17
3020	2x8	"	2.33	1.21	3.54
3030	2x10	"	2.42	1.62	4.04
3040	2x12	"	2.52	2.09	4.61
3180	Jack rafters				
3190	4-6 pitch (4-in-12 to 6-in-12)				
3200	16" o.c.				
3210	2x6	S.F.	1.85	1.13	2.98
3220	2x8	"	1.90	1.74	3.64
3230	2x10	"	2.03	1.99	4.02
3240	2x12	"	2.10	2.96	5.06
3250	24" o.c.				
3260	2x6	S.F.	1.43	0.87	2.30
3270	2x8	"	1.46	1.48	2.94
3280	2x10	"	1.53	1.74	3.27
3290	2x12	"	1.57	2.52	4.09
3295	8-12 pitch (8-in-12 to 12-in-12)				
3300	16" o.c.				
3310	2x6	S.F.	1.96	1.74	3.70
3320	2x8	"	2.03	2.17	4.20

		UNIT	LABOR	MAT.	TOTAL
06110.50	**ROOF FRAMING, Cont'd...**				
3330	2x10	S.F.	2.10	3.13	5.23
3340	2x12	"	2.17	4.34	6.51
3350	24" o.c.				
3360	2x6	S.F.	1.50	1.38	2.88
3370	2x8	"	1.53	1.74	3.27
3380	2x10	"	1.57	2.78	4.35
3390	2x12	"	1.61	4.00	5.61
4980	Sister rafters				
5000	2x4	L.F.	4.50	0.60	5.10
5010	2x6	"	5.25	0.92	6.17
5020	2x8	"	6.30	1.21	7.51
5030	2x10	"	7.87	1.62	9.49
5040	2x12	"	10.50	2.09	12.59
5050	Fascia boards				
5060	2x4	L.F.	3.15	0.60	3.75
5070	2x6	"	3.15	0.92	4.07
5080	2x8	"	3.50	1.21	4.71
5090	2x10	"	3.50	1.62	5.12
5100	2x12	"	3.93	2.09	6.02
7980	Cant strips				
7985	Fiber				
8000	3x3	L.F.	1.80	0.48	2.28
8020	4x4	"	1.90	0.67	2.57
8030	Wood				
8040	3x3	L.F.	1.90	2.52	4.42
06110.60	**SLEEPERS**				
0960	Sleepers, over concrete				
1090	16" o.c.				
1100	1x2	S.F.	1.26	0.26	1.52
1120	1x3	"	1.26	0.37	1.63
1140	2x4	"	1.50	0.79	2.29
1160	2x6	"	1.57	1.17	2.74
06110.65	**SOFFITS**				
0980	Soffit framing				
1000	2x3	L.F.	4.50	0.41	4.91
1020	2x4	"	4.84	0.51	5.35
1030	2x6	"	5.25	0.75	6.00
1040	2x8	"	5.72	1.06	6.78
06110.70	**WALL FRAMING**				
0960	Framing wall, studs				
1110	16" o.c.				
1120	2x3	S.F.	0.98	0.43	1.41
1140	2x4	"	0.98	0.61	1.59
1150	2x6	"	1.05	0.87	1.92
1160	2x8	"	1.08	1.37	2.45
1165	24" o.c.				
1170	2x3	S.F.	0.85	0.34	1.19
1180	2x4	"	0.85	0.46	1.31
1190	2x6	"	0.90	0.73	1.63
1200	2x8	"	0.92	0.95	1.87
1480	Plates, top or bottom				

		UNIT	LABOR	MAT.	TOTAL
06110.70	**WALL FRAMING, Cont'd...**				
1500	2x3	L.F.	1.85	0.41	2.26
1510	2x4	"	1.96	0.51	2.47
1520	2x6	"	2.10	0.75	2.85
1530	2x8	"	2.25	1.06	3.31
2000	Headers, door or window				
2044	2x8				
2046	Single				
2050	4' long	EA.	39.25	4.51	43.76
2060	8' long	"	48.50	9.01	57.51
2065	Double				
2070	4' long	EA.	45.00	9.01	54.01
2080	8' long	"	57.00	18.00	75.00
2134	2x12				
2138	Single				
2140	6' long	EA.	48.50	9.89	58.39
2150	12' long	"	63.00	19.50	82.50
2155	Double				
2160	6' long	EA.	57.00	19.50	76.50
2170	12' long	"	70.00	38.75	109
06115.10	**FLOOR SHEATHING**				
1980	Sub-flooring, plywood, CDX				
2000	1/2" thick	S.F.	0.78	0.61	1.39
2020	5/8" thick	"	0.90	0.88	1.78
2080	3/4" thick	"	1.05	1.62	2.67
2090	Structural plywood				
2100	1/2" thick	S.F.	0.78	0.96	1.74
2120	5/8" thick	"	0.90	1.54	2.44
2140	3/4" thick	"	0.96	1.62	2.58
5990	Underlayment				
6000	Hardboard, 1/4" tempered	S.F.	0.78	0.90	1.68
6010	Plywood, CDX				
6020	3/8" thick	S.F.	0.78	0.94	1.72
6040	1/2" thick	"	0.84	1.12	1.96
6060	5/8" thick	"	0.90	1.30	2.20
6080	3/4" thick	"	0.96	1.62	2.58
06115.20	**ROOF SHEATHING**				
0080	Sheathing				
0090	Plywood, CDX				
1000	3/8" thick	S.F.	0.81	0.94	1.75
1020	1/2" thick	"	0.84	1.12	1.96
1040	5/8" thick	"	0.90	1.30	2.20
1060	3/4" thick	"	0.96	1.62	2.58
1080	Structural plywood				
2040	3/8" thick	S.F.	0.81	0.59	1.40
2060	1/2" thick	"	0.84	0.77	1.61
2080	5/8" thick	"	0.90	0.94	1.84
2100	3/4" thick	"	0.96	1.13	2.09
06115.30	**WALL SHEATHING**				
0980	Sheathing				
0990	Plywood, CDX				
1000	3/8" thick	S.F.	0.93	0.94	1.87

		UNIT	LABOR	MAT.	TOTAL
06115.30	**WALL SHEATHING, Cont'd...**				
1020	1/2" thick	S.F.	0.96	1.12	2.08
1040	5/8" thick	"	1.05	1.30	2.35
1060	3/4" thick	"	1.14	1.62	2.76
3000	Waferboard				
3020	3/8" thick	S.F.	0.93	0.59	1.52
3040	1/2" thick	"	0.96	0.77	1.73
3060	5/8" thick	"	1.05	0.94	1.99
3080	3/4" thick	"	1.14	1.03	2.17
4100	Structural plywood				
4120	3/8" thick	S.F.	0.93	0.94	1.87
4140	1/2" thick	"	0.96	1.12	2.08
4160	5/8" thick	"	1.05	1.30	2.35
4180	3/4" thick	"	1.14	1.12	2.26
7000	Gypsum, 1/2" thick	"	0.96	0.59	1.55
8000	Asphalt impregnated fiberboard, 1/2" thick	"	0.96	1.03	1.99
06125.10	**WOOD DECKING**				
0090	Decking, T&G solid				
0095	Cedar				
0100	3" thick	S.F.	1.57	11.75	13.32
0120	4" thick	"	1.68	14.50	16.18
1030	Fir				
1040	3" thick	S.F.	1.57	5.10	6.67
1060	4" thick	"	1.68	6.19	7.87
1080	Southern yellow pine				
2000	3" thick	S.F.	1.80	5.10	6.90
2020	4" thick	"	1.93	5.39	7.32
3120	White pine				
3140	3" thick	S.F.	1.57	6.19	7.76
3160	4" thick	"	1.68	8.38	10.06
06130.10	**HEAVY TIMBER**				
1000	Mill framed structures				
1010	Beams to 20' long				
1020	Douglas fir				
1040	6x8	L.F.	8.26	8.36	16.62
1042	6x10	"	8.54	9.87	18.41
1044	6x12	"	9.17	11.75	20.92
1046	6x14	"	9.53	14.25	23.78
1048	6x16	"	9.91	15.50	25.41
1060	8x10	"	8.54	13.00	21.54
1070	8x12	"	9.17	15.50	24.67
1080	8x14	"	9.53	17.75	27.28
1090	8x16	"	9.91	20.25	30.16
1380	Columns to 12' high				
1400	Douglas fir				
1420	6x6	L.F.	12.50	6.01	18.51
1440	8x8	"	12.50	10.25	22.75
1460	10x10	"	13.75	18.00	31.75
1480	12x12	"	13.75	22.25	36.00
2000	Posts, treated				
2100	4x4	L.F.	2.52	2.07	4.59
2120	6x6	"	3.15	6.01	9.16

		UNIT	LABOR	MAT.	TOTAL
06190.20	**WOOD TRUSSES**				
0960	Truss, fink, 2x4 members				
0980	3-in-12 slope				
1030	5-in-12 slope				
1040	24' span	EA.	73.00	130	203
1050	28' span	"	75.00	140	215
1055	30' span	"	77.00	150	227
1060	32' span	"	77.00	160	237
1070	40' span	"	83.00	210	293
1074	Gable, 2x4 members				
1078	5-in-12 slope				
1080	24' span	EA.	73.00	150	223
1100	28' span	"	75.00	180	255
1120	30' span	"	77.00	190	267
1160	36' span	"	80.00	210	290
1180	40' span	"	83.00	230	313
1190	King post type, 2x4 members				
2000	4-in-12 slope				
2040	16' span	EA.	67.00	91.00	158
2060	18' span	"	69.00	98.00	167
2080	24' span	"	73.00	110	183
2120	30' span	"	77.00	140	217
2160	38' span	"	80.00	180	260
2180	42' span	"	85.00	220	305
06190.30	**LAMINATED BEAMS**				
0010	Parallel strand beams 3-1/2" wide x				
0020	9-1/2"	L.F.	3.54	12.25	15.79
0030	11-1/4"	"	3.67	13.00	16.67
0040	11-7/8"	"	3.81	13.75	17.56
0050	14"	"	4.50	17.50	22.00
0060	16"	"	4.95	20.75	25.70
0070	18"	"	5.50	24.50	30.00
1000	Laminated veneer beams, 1-3/4" wide x				
1010	11-7/8"	L.F.	3.81	8.36	12.17
1020	14"	"	4.50	10.50	15.00
1030	16"	"	4.95	10.25	15.20
1040	18"	"	5.50	13.00	18.50
2000	Laminated strand beams, 1-3/4" wide x				
2010	9-1/2"	L.F.	3.54	5.36	8.90
2020	11-7/8"	"	3.81	6.13	9.94
2030	14"	"	4.50	7.21	11.71
2040	16"	"	4.95	8.27	13.22
2050	3-1/2" wide x				
2060	9-1/2"	L.F.	3.54	9.43	12.97
2070	11-7/8"	"	3.81	12.25	16.06
2080	14"	"	4.50	14.50	19.00
2090	16"	"	4.95	17.25	22.20
3000	Gluelam beam, 3-1/2" wide x				
3010	10"	L.F.	3.54	14.25	17.79
3020	12"	"	4.13	16.75	20.88
3030	15"	"	4.72	20.00	24.72
3040	5-1/2" wide x				

		UNIT	LABOR	MAT.	TOTAL
06190.30	**LAMINATED BEAMS, Cont'd...**				
3050	10"	L.F.	3.54	23.00	26.54
3060	16"	"	4.95	36.00	40.95
3070	20"	"	5.83	42.50	48.33
06200.10	**FINISH CARPENTRY**				
0070	Mouldings and trim				
0980	Apron, flat				
1000	9/16 x 2	L.F.	3.15	1.99	5.14
1010	9/16 x 3-1/2	"	3.31	4.59	7.90
1015	Base				
1020	Colonial				
1022	7/16 x 2-1/4	L.F.	3.15	2.37	5.52
1024	7/16 x 3	"	3.15	3.07	6.22
1026	7/16 x 3-1/4	"	3.15	3.14	6.29
1028	9/16 x 3	"	3.31	3.07	6.38
1030	9/16 x 3-1/4	"	3.31	3.21	6.52
1034	11/16 x 2-1/4	"	3.50	3.37	6.87
1035	Ranch				
1036	7/16 x 2-1/4	L.F.	3.15	2.60	5.75
1038	7/16 x 3-1/4	"	3.15	3.07	6.22
1039	9/16 x 2-1/4	"	3.31	2.83	6.14
1041	9/16 x 3	"	3.31	3.07	6.38
1043	9/16 x 3-1/4	"	3.31	3.14	6.45
1050	Casing				
1060	11/16 x 2-1/2	L.F.	2.86	2.44	5.30
1070	11/16 x 3-1/2	"	3.00	2.76	5.76
1180	Chair rail				
1200	9/16 x 2-1/2	L.F.	3.15	2.60	5.75
1210	9/16 x 3-1/2	"	3.15	3.60	6.75
1250	Closet pole				
1300	1-1/8" dia.	L.F.	4.20	1.76	5.96
1310	1-5/8" dia.	"	4.20	2.60	6.80
1340	Cove				
1500	9/16 x 1-3/4	L.F.	3.15	1.99	5.14
1510	11/16 x 2-3/4	"	3.15	3.07	6.22
1550	Crown				
1600	9/16 x 1-5/8	L.F.	4.20	2.60	6.80
1620	11/16 x 3-5/8	"	5.25	3.07	8.32
1640	11/16 x 5-1/4	"	6.30	5.14	11.44
1680	Drip cap				
1700	1-1/16 x 1-5/8	L.F.	3.15	2.76	5.91
1780	Glass bead				
1800	3/8 x 3/8	L.F.	3.93	0.99	4.92
1840	5/8 x 5/8	"	3.93	1.30	5.23
1860	3/4 x 3/4	"	3.93	1.53	5.46
1880	Half round				
1900	1/2	L.F.	2.52	1.15	3.67
1910	5/8	"	2.52	1.53	4.05
1920	3/4	"	2.52	2.07	4.59
1980	Lattice				
2000	1/4 x 7/8	L.F.	2.52	0.92	3.44
2010	1/4 x 1-1/8	"	2.52	0.99	3.51

		UNIT	LABOR	MAT.	TOTAL
06200.10	**FINISH CARPENTRY, Cont'd...**				
2030	1/4 x 1-3/4	L.F.	2.52	1.19	3.71
2040	1/4 x 2	"	2.52	1.38	3.90
2080	Ogee molding				
2100	5/8 x 3/4	L.F.	3.15	1.83	4.98
2110	11/16 x 1-1/8	"	3.15	4.30	7.45
2120	11/16 x 1-3/8	"	3.15	3.37	6.52
2180	Parting bead				
2200	3/8 x 7/8	L.F.	3.93	1.53	5.46
2300	Quarter round				
2301	1/4 x 1/4	L.F.	2.52	0.54	3.06
2303	3/8 x 3/8	"	2.52	0.76	3.28
2305	1/2 x 1/2	"	2.52	0.99	3.51
2307	11/16 x 11/16	"	2.73	0.99	3.72
2309	3/4 x 3/4	"	2.73	1.83	4.56
2311	1-1/16 x 1-1/16	"	2.86	1.45	4.31
2380	Railings, balusters				
2400	1-1/8 x 1-1/8	L.F.	6.30	4.91	11.21
2410	1-1/2 x 1-1/2	"	5.72	5.75	11.47
2480	Screen moldings				
2500	1/4 x 3/4	L.F.	5.25	1.22	6.47
2510	5/8 x 5/16	"	5.25	1.53	6.78
2580	Shoe				
2600	7/16 x 11/16	L.F.	2.52	1.53	4.05
2605	Sash beads				
2620	1/2 x 7/8	L.F.	5.25	1.99	7.24
2640	5/8 x 7/8	"	5.72	2.15	7.87
2760	Stop				
2780	5/8 x 1-5/8				
2800	Colonial	L.F.	3.93	1.06	4.99
2810	Ranch	"	3.93	1.06	4.99
2880	Stools				
2900	11/16 x 2-1/4	L.F.	7.00	4.68	11.68
2910	11/16 x 2-1/2	"	7.00	4.91	11.91
2920	11/16 x 5-1/4	"	7.87	5.06	12.93
4000	Exterior trim, casing, select pine, 1x3	"	3.15	3.37	6.52
4010	Douglas fir				
4020	1x3	L.F.	3.15	1.60	4.75
4040	1x4	"	3.15	1.99	5.14
4060	1x6	"	3.50	2.60	6.10
4100	1x8	"	3.93	3.60	7.53
5000	Cornices, white pine, #2 or better				
5040	1x4	L.F.	3.15	1.22	4.37
5080	1x8	"	3.70	2.44	6.14
5120	1x12	"	4.20	3.91	8.11
8600	Shelving, pine				
8620	1x8	L.F.	4.84	1.76	6.60
8640	1x10	"	5.04	2.30	7.34
8660	1x12	"	5.25	2.91	8.16
8800	Plywood shelf, 3/4", with edge band, 12" wide	"	6.30	3.14	9.44
8840	Adjustable shelf, and rod, 12" wide				
8860	3' to 4' long	EA.	15.75	25.00	40.75
8880	5' to 8' long	"	21.00	47.00	68.00

		UNIT	LABOR	MAT.	TOTAL
06200.10	**FINISH CARPENTRY, Cont'd...**				
8900	Prefinished wood shelves with brackets and supports				
8905	8" wide				
8910	3' long	EA.	15.75	74.00	89.75
8922	4' long	"	15.75	85.00	101
8924	6' long	"	15.75	120	136
8930	10" wide				
8940	3' long	EA.	15.75	81.00	96.75
8942	4' long	"	15.75	120	136
8946	6' long	"	15.75	130	146
06220.10	**MILLWORK**				
0070	Countertop, laminated plastic				
0080	25" x 7/8" thick				
0099	Minimum	L.F.	15.75	18.00	33.75
0100	Average	"	21.00	34.25	55.25
0110	Maximum	"	25.25	50.00	75.25
0115	25" x 1-1/4" thick				
0120	Minimum	L.F.	21.00	22.00	43.00
0130	Average	"	25.25	43.75	69.00
0140	Maximum	"	31.50	66.00	97.50
0160	Add for cutouts	EA.	39.25		39.25
0165	Backsplash, 4" high, 7/8" thick	L.F.	12.50	24.00	36.50
2000	Plywood, sanded, A-C				
2020	1/4" thick	S.F.	2.10	1.55	3.65
2040	3/8" thick	"	2.25	1.69	3.94
2060	1/2" thick	"	2.42	1.91	4.33
2070	A-D				
2080	1/4" thick	S.F.	2.10	1.48	3.58
2090	3/8" thick	"	2.25	1.69	3.94
2100	1/2" thick	"	2.42	1.83	4.25
2500	Base cabinet, 34-1/2" high, 24" deep, hardwood				
2540	Minimum	L.F.	25.25	240	265
2560	Average	"	31.50	270	302
2580	Maximum	"	42.00	300	342
2600	Wall cabinets				
2640	Minimum	L.F.	21.00	72.00	93.00
2660	Average	"	25.25	97.00	122
2680	Maximum	"	31.50	120	152
06300.10	**WOOD TREATMENT**				
1000	Creosote preservative treatment				
1020	8 lb/cf	B.F.			0.74
1040	10 lb/cf	"			0.89
1060	Salt preservative treatment				
1070	Oil borne				
1080	Minimum	B.F.			0.68
1100	Maximum	"			0.96
1120	Water borne				
1140	Minimum	B.F.			0.48
1150	Maximum	"			0.74
1200	Fire retardant treatment				
1220	Minimum	B.F.			0.96
1240	Maximum	"			1.16

		UNIT	LABOR	MAT.	TOTAL
06300.10	**WOOD TREATMENT, Cont'd...**				
1300	Kiln dried, softwood, add to framing costs				
1320	1" thick	B.F.			0.34
1360	3" thick	"			0.61
06430.10	**STAIRWORK**				
0080	Risers, 1x8, 42" wide				
0100	White oak	EA.	31.50	51.00	82.50
0120	Pine	"	31.50	45.25	76.75
0130	Treads, 1-1/16" x 9-1/2" x 42"				
0140	White oak	EA.	39.25	61.00	100
06440.10	**COLUMNS**				
0980	Column, hollow, round wood				
0990	12" diameter				
1000	10' high	EA.	91.00	910	1,001
1080	16' high	"	140	1,650	1,790
2000	24" diameter				
2020	16' high	EA.	140	3,770	3,910
2060	20' high	"	140	5,260	5,400
2100	24' high	"	150	6,050	6,200

		UNIT	COST
06999.10	**ROUGH CARPENTRY**		
1000	LIGHT FRAMING AND SHEATHING		
1100	Joists and Headers - Floor Area - 16" O.C.		
1110	2" x 6" Joists - with Headers & Bridging	S.F.	3.52
1120	2" x 8" Joists	"	4.34
1130	2" x 10" Joists	"	4.83
1140	2" x 12" Joists	"	5.76
1150	Add for Ceiling Joists, 2nd Floor and Above	"	0.13
1160	Add for Sloped Installation	"	0.23
1200	Studs, Plates and Framing - 8' Wall Height - 16" O.C.	"	2.12
1210	2" x 3" Stud Wall - Non-Bearing (Single Top Plate)		
1220	2" x 4" Stud Wall - Bearing (Double Top Plate)	S.F.	2.53
1230	2" x 4" Stud Wall - Non-Bearing (Single Top Plate)	"	1.87
1240	2" x 6" Stud Wall - Bearing (Double Top Plate)	"	3.43
1250	2" x 6" Stud Wall - Non-Bearing (Single Top Plate)	"	2.55
1300	Add for Stud Wall - 12" O.C.	"	0.12
1310	Deduct for Stud Wall - 24" O.C.	"	0.31
1320	Add for Bolted Plates or Sills	"	0.18
1330	Add for Each Foot Above 8'	"	0.10
1340	Add to Above for Fire Stops, Fillers and Nailers	"	0.67
1350	Add for Soffits and Suspended Framing	"	0.92
1360	Bridging - 1" x 3" Wood Diagonal	EA.	3.57
1370	2" x 8" Solid	"	3.97
1400	Rafters		
1410	2" x 4" Rafter (Incl Bracing) 3 - 12 Slope	S.F.	2.60
1420	4 - 12 Slope	"	2.60
1430	5 - 12 Slope	"	2.94
1440	6 - 12 Slope	"	3.45
1500	Add for Hip-and-Valley Type	"	0.69
1510	Add for 1' 0" Overhang - Total Area of Roof	"	0.26
1520	2" x 6" Rafter (Incl Bracing) 3-12 Slope	"	3.29
1530	4-12 Slope	"	3.32
1540	5-12 Slope	"	3.41
1550	6-12 Slope	"	3.60
1560	Add for Hip - and - Valley Type	"	0.77
1600	Stairs	EA.	660
1700	Sub Floor Sheathing (Structural)		
1710	1" x 8" and 1" x 10" #3 Pine	S.F.	2.32
1720	1/2" x 4' x 8' CD Plywood - Exterior	"	1.85
1730	5/8" x 4' x 8'	"	2.25
1740	3/4" x 4' x 8'	"	3.01
1800	Floor Sheathing (Over Sub Floor)		
1810	3/8" x 4' x 8' CD Plywood	S.F.	1.72
1820	1/2" x 4' x 8'	"	2.07
1830	5/8" x 4' x 8'	"	2.71
1840	1/2" x 4' x 8' Particle Board	"	2.14
1850	5/8" x 4' x 8'	"	2.50
1860	3/4" x 4' x 8'	"	3.08
1900	Wall Sheathing		
1910	1' x 8" and 1" x 10" - #3 Pine	S.F.	2.37
1920	3/8" x 4' x 8' CD Plywood - Exterior	"	1.77
1930	1/2" x 4' x 8'	"	2.07
1940	5/8" x 4' x 8'	"	2.83

		UNIT	COST
06999.10	**ROUGH CARPENTRY, Cont'd...**		
1950	3/4" x 4' x 8'	S.F.	3.18
1960	25/32" x 4' x 8' Fiber Board - Impregnated	"	1.98
1970	1" x 2' x 8' T&G Styrofoam	"	1.75
1980	2" x 2' x 8'	"	2.56
2000	Roof Sheathing - Flat Construction		
2010	1" x 6" and 1" x 8" - #3 Pine	S.F.	2.93
2020	1/2" x 4' x 8' CD Plywood - Exterior	"	2.73
2030	5/8" x 4' x 8'	"	2.38
2040	3/4" x 4' x 8'	"	3.19
2050	Add for Sloped Roof Construction (to 5-12 slope)	"	0.27
2060	Add for Steep Sloped Construction (over 5-12 slope)	"	0.49
2070	Add to Above Sheathing		
2080	AC or AD Plywood	S.F.	0.49
2090	10' Length Plywood	"	0.37
2100	HEAVY FRAMING		
2200	Columns and Beams - 16' Span Average - Floor Area	S.F.	10.12
2210	20' Span	"	10.53
2220	24' Span	"	12.42
2300	Deck 2" x 6" T&G - Fir Random Construction Grade	"	6.14
2310	3" x 6" T&G - Fir Random Construction Grade	"	8.17
2320	4" x 6"	"	10.12
2330	2" x 6" T&G - Red Cedar	"	6.46
2340	Add for D Grade Cedar	"	2.53
2350	2" x 6" T&G - Panelized Fir	"	5.46
3000	MISCELLANEOUS CARPENTRY		
3100	Blocking and Bucks (2" x 4" and 2" x 6")		
3110	Doors & Windows - Nailed to Concrete or Masonry	EA.	73.75
3120	Bolted to Concrete or Masonry	"	81.50
3130	Doors & Windows - Nailed to Concrete or Masonry	B.F.	3.57
3140	Bolted to Concrete or Masonry	"	4.94
3150	Roof Edges - Nailed to Wood	"	2.87
3160	Bolted to Concrete (Incl. bolts)	"	4.28
3200	Grounds and Furring		
3210	1" x 6" Fastened to Wood	L.F.	2.27
3220	1" x 4"	"	1.70
3230	1" x 3"	"	1.54
3240	1" x 3" Fastened to Concrete/Masonry - Nailed	"	2.37
3250	Gun Driven	"	2.15
3260	PreClipped	"	2.69
3270	2" x 2" Suspended Framing	"	2.37
3280	Not Suspended Framing	"	2.05
3300	Grounds and Furring		
3310	1" x 6" Fastened to Wood	B.F.	4.57
3320	1" x 4"	"	5.39
3330	1" x 3"	"	6.04
3340	1" x 3" Fastened to Concrete/Masonry - Nailed	"	9.02
3400	Gun Driven	"	8.57
3410	PreClipped	"	10.51
3420	2" x 2" Suspended Framing	"	7.04
3430	Not Suspended Framing	"	6.09
3440	Cant Strips		
3450	4" x 4" Treated and Nailed	S.F.	2.70

		UNIT	COST
06999.10	**ROUGH CARPENTRY, Cont'd...**		
3460	Treated and Bolted (Including Bolts)	S.F.	3.91
3470	6" x 6" Treated and Nailed	"	5.41
3480	Treated and Bolted (Including Bolts)	"	6.87
3600	Building Papers and Sealers		
3610	15" Felt	S.F.	0.38
3620	Polyethylene - 4 mil	"	0.31
3630	6 mil	"	0.35
3640	Sill Sealer	"	1.34
06999.20	**FINISH CARPENTRY**		
1000	FINISH SIDINGS AND FACING MATERIALS (Exterior) TO %		
1100	Boards and Beveled Sidings		
1200	Cedar Beveled - Clear Heart 1/2" x 4"	S.F.	8.43
1210	1/2" x 6"	"	7.43
1220	1/2" x 8"	"	6.36
1300	Rough Sawn 7/8" x 8"	"	6.48
1310	7/8" x 10"	"	6.69
1320	7/8" x 12"	"	6.69
1330	3/4" x 12"	"	5.83
1400	Redwood Beveled - Clear Heart		
1410	1/2" x 6"	S.F.	8.10
1430	5/8" x 10"	"	8.24
1440	3/4" x 6"	"	8.68
1450	3/4" x 8"	"	9.49
1500	3/4" x 6" Board - Clear	"	4.87
1510	3/4" Tongue & Groove	"	7.83
1520	1" Rustic Beveled - 6" and 8"	"	8.46
1530	5/4" Rustic Beveled - 6" and 8"	"	8.46
1540	Add for Metal Corners	EA.	2.35
1550	Add for Mitering Corners	"	4.94
1600	Plywood Fir AC Smooth -One Side 1/4"	S.F.	2.96
1610	3/8"	"	3.15
1620	1/2"	"	3.20
1630	5/8"	"	3.39
1640	3/4"	"	3.79
1650	5/8" - Grooved and Rough Faced	"	3.90
1700	Cedar - Rough Sawn 3/8"	"	3.79
1710	5/8"	"	3.90
1720	3/4"	"	4.62
1730	5/8" - Grooved and Rough Faced	"	4.62
1740	Add for Wood Batten Strips, 1" x 2" - 4' O.C.	"	0.62
1750	Add for Splines	"	0.62
1800	Hardboard - Paneling and Lap Siding (Primed)		
1810	3/8" Rough Textured Paneling - 4' x 8'	S.F.	3.13
1820	7/16" Grooved - 4' x 8'	"	2.96
1830	7/16" Stucco Board Paneling	"	3.27
1840	7/16" x 8" Lap Siding	"	3.75
1860	Add for Pre-Finishing	"	0.31
1900	Shingles		
1910	Wood - 16" Red Cedar - 12" to Weather - #1	S.F.	6.16
1920	#2	"	4.86
1930	#3	"	4.56

		UNIT	COST
06999.20	**FINISH CARPENTRY, Cont'd...**		
1940	Red Cedar Hand Splits - #1, 24" - 1/2" to 3/4"	S.F.	5.17
1950	24" - 3/4" to 1 1/4"	"	5.76
1960	#2	"	5.17
1970	#3	"	4.86
2000	Add for Fire Retardant	"	0.74
2010	Add for 3/8" Backer Board	"	1.23
2020	Add for Metal Corners	EA.	2.38
2030	Add for Ridges, Hips and Corners	L.F.	6.36
2040	Add for 8" to Weather	PCT.	29.50
2100	Facia - Pine #2 1" x 8"	L.F.	3.01
2110	Cedar #3 1" x 8"	"	4.77
2120	Redwood - Clear 1" x 8"	"	6.34
2130	Plywood 5/8" - ACX 1" x 8"	"	3.11
2140	Facia - Pine #2 1" x 8"	S.F.	4.50
2200	Cedar #3 1" x 8"	"	6.67
2210	Redwood - Clear 1" x 8"	"	9.70
2220	Plywood 5/8" - ACX 1" x 8"	"	4.58
2300	FINISH WALLS (Interior) (15% Added for Waste)	"	5.01
2400	Boards, Cedar - #3 1" x 6" and 1" x 8"	"	5.01
2410	Knotty 1" x 6" and 1" x 8"	"	5.32
2420	D Grade 1" x 6" and 1" x 8"	"	6.44
2430	Aromatic 1" x 6" and 1" x 8"	"	6.64
2440	Redwood - Construction 1" x 6" and 1" x 8"	"	7.97
2450	Clear 1" x 6" and 1" x 8"	"	8.25
2460	Fir - Beaded 5/8" x 4"	"	7.08
2470	Pine - #2 1" x 6" and 1" x 8"	"	8.92
2500	Hardboard (Paneling) Tempered - 1/8"	"	6.54
2510	1/4"	"	4.98
2600	Plywood (Prefinished Paneling) 1/4" Birch - Natural	"	2.08
2610	3/4" Birch - Natural	"	2.28
2620	1/4" Birch - White	"	4.33
2630	1/4" Oak - Rotary Cut	"	5.71
2640	3/4"	"	7.58
2700	1/4" Oak - White	"	5.55
2710	1/4" Mahogany (Lauan)	"	6.88
2720	3/4"	"	9.34
2730	3/4" Mahogany (African)	"	4.38
2740	1/4" Walnut	"	6.26
2800	Gypsum Board (Paneling)	"	7.51
2810	Prefinished	"	8.25
2820	Red Oak Plastic		
06999.30	**MILLWORK & CUSTOM WOODWORK**		
1000	CUSTOM CABINET WORK (Red Oak or Birch)		
1010	Base Cabinets - Avg. 35" H x 24" D with Drawer	L.F.	440
1020	Sink Fronts	"	171
1030	Corner Cabinets	"	350
1040	Add per Drawer	"	73.75
1050	Upper Cabinets - Avg. 30" High x 12" Deep	"	220
2000	Utility Cabinets - Avg. 84" High x 24" Deep	"	520
2010	China or Corner Cabinets - 84" High	EA.	1,000
2020	Oven Cabinets - 84" High x 24" Deep	"	590

		UNIT	COST
06999.30	**MILLWORK & CUSTOM WOODWORK, Cont'd...**		
2030	Vanity Cabinets - Avg. 30" High x 21" Deep	EA.	350
2040	Deduct for Prefinishing wood	PCT.	10.42
2050	Base Cabinets - Avg. 35" H x 24" D with Drawer	EA.	270
2060	Sink Fronts	L.F.	171
2070	Corner Cabinets	"	380
2080	Add per Drawer	"	75.25
2090	Upper Cabinets - Avg. 30" High x 12" Deep	"	220
3000	Utility Cabinets - Avg. 84" High x 24" Deep	"	520
3010	China or Corner Cabinets - 84" High	"	1,010
3020	Oven Cabinets - 84" High x 24" Deep	"	580
3030	Vanity Cabinets - Avg. 30" High x 21" Deep	"	350
4000	COUNTER TOPS - 25"		
4010	Plastic Laminated with 4" Back Splash	L.F.	101
4020	Deduct for No Back Splash	"	16.95
4030	Granite - 1 1/4" - Artificial	"	181
4040	3/4" - Artificial	"	142
4050	Marble	"	172
4070	Wood Cutting Block	"	172
5000	Stainless Steel	"	250
5010	Polyester-Acrylic Solid Surface	"	290
5020	Quartz	"	156
5030	Plastic (Polymer)	"	106
6000	CUSTOM DOOR FRAMES - Including 2 Sides Trim		
6010	Birch	EA.	450
6020	Fir	"	280
6030	Poplar	"	280
6040	Oak	"	500
6050	Pine	"	370
6060	Walnut	"	590
8000	MOULDINGS AND TRIM		
8010	Apron 7/16" x 2"	L.F.	3.52
8020	Astragal 1 3/4" x 2 1/4"	"	8.34
8030	Base 7/16" x 2 3/4"	"	5.04
8050	Base Shoe 7/16" x 2 3/4"	"	4.03
8060	Batten Strip 5/8" x 1 5/8"	"	3.48
8070	Brick Mould 1 1/14" x 2"	"	4.05
9080	Casing 11/16" x 2 1/4"	"	4.38
9100	Chair Rail 5/8" x 1 3/4"	"	4.60
9110	Closet Rod 1 5/16"	"	4.79
9120	Corner Bead 1 1/8" x 1 1/8"	"	6.06
9130	Cove Moulding 3/4" x 3/4"	"	3.78
9140	Crown Moulding 9/16" x 3 5/8"	"	5.62
9160	Drip Cap	"	4.53
9170	Half Round 1/2" x 1"	"	4.84
9180	Hand Rail 1 5/8" x 1 3/4"	"	6.79
9190	Hook Strip 5/8" x 2 1/2"	"	3.89
9200	Picture Mould 3/4" x 1 1/2"	"	4.33
9210	Quarter Round 3/4" x 3/4"	"	3.08
9220	1/2" x 1/2"	"	2.36
9230	Sill 3/4" x 2 1/2"	"	5.17
9240	Stool 11/16 x 2 1/2"	"	6.21
9250	BIRCH		

		UNIT	COST
06999.30	**MILLWORK & CUSTOM WOODWORK, Cont'd...**		
9260	STAIRS (Treads, Risers, Skirt Boards)	L.F.	59.00
9270	SHELVING (12" Deep)	S.F.	62.50
9280	CUSTOM PANELING	"	42.75
9290	OAK		
9300	STAIRS (Treads, Risers, Skirt Boards)	L.F.	46.00
9310	SHELVING (12" Deep)	S.F.	55.75
9320	CUSTOM PANELING	"	32.00
9330	THRESHOLDS - 3/4" x 3 1/2"	L.F.	30.25
9340	PINE		
9350	STAIRS (Treads, Risers, Skirt Boards)	L.F.	34.50
9360	SHELVING (12" Deep)	"	38.50
9370	CUSTOM PANELING	S.F.	26.25
06999.40	**GLUE LAMINATE**		
0010	Arches 80' Span	S.F.	15.27
0020	Beams & Purlins 40' Span	"	10.58
0030	Deck Fir - 3" x 6"	"	12.60
0040	4" x 6"	"	13.54
0050	Deck Cedar - 3" x 6"	"	16.33
0060	4" x 6"	"	18.41
0070	Deck Pine - 3" x 6"	"	11.04
06999.50	**PREFABRICATED WOOD COMPONENTS**		
1000	WOOD TRUSSED RAFTERS		
1010	24" O.C. - 4 -12 Pitch		
2000	Span To 16' w/ Supports	EA.	210
2010	20'	"	220
2030	24'	"	260
2050	28'	"	270
2070	32'	"	320
3000	Add for 5-12 Pitch	PCT.	5.21
3010	Add for 6-12 Pitch	"	10.42
3020	Add for Scissor Truss	EA.	45.50
3030	Add for Gable End 24'	"	97.75
3040	Span To 16' w/ Supports	S.F.	6.45
3050	20'	"	5.23
3070	24'	"	4.95
3100	28'	"	4.99
3120	32'	"	5.58
5000	Add for 5-12 Pitch	PCT.	15.63
5010	Add for 6-12 Pitch	"	26.05
5020	Add for Scissor Truss	S.F.	36.00
5030	Add for Gable End 24'	"	0.65
6000	WOOD FLOOR TRUSS JOISTS (TJI)		
6010	Open Web - Wood or Metal - 24" O.C.		
6020	Up To 23' Span x 12" Single	L.F.	12.92
6030	To 24' Span x 15" Cord	"	13.24
6050	To 30' Span x 21"	"	14.43
6060	To 25' Span x 15" Double	"	12.57
6070	To 30' Span x 18" Cord	"	13.31
6080	To 36' Span x 21"	"	14.87
7000	Plywood Web - 24" O.C.		
7010	Up To 15' x 9 1/2"	L.F.	8.35

		UNIT	COST
06999.50	**PREFABRICATED WOOD COMPONENTS, Cont'd...**		
7030	To 21' x 14"	L.F.	9.28
7040	To 22' x 16"	"	10.48
8000	Open Web - Wood or Metal - 24" O.C.		
8010	Up To 23' Span x 12" Single	S.F.	6.44
8020	To 24' Span x 15" Cord	"	5.22
8030	To 27' Span x 18"	"	6.84
8040	To 30' Span x 21"	"	7.26
8050	To 25' Span x 15" Double	"	6.35
8060	To 30' Span x 18" Cord	"	6.49
8070	To 36' Span x 21"	"	7.34
8500	Plywood Web - 24" O.C.		
8510	Up To 15' x 9 1/2"	S.F.	4.26
8520	To 19' x 11 7/8"	"	4.25
8530	To 21' x 14"	"	4.75
8540	To 22' x 16"	"	5.23
06999.51	**LAMINATED VENEER STRUCTURAL BEAMS**		
0005	Micro Lam - Plywood - 24" O.C.		
0010	9 1/2" x 1 3/4"	L.F.	12.65
0020	11 7/8" x 1 3/4"	"	13.70
0030	14" x 1 3/4"	"	16.27
0040	16" x 1 3/4"	"	17.98
1000	Glue Lam- Dimension Lumber- 24" O.C.		
1010	9" x 3 1/2"	L.F.	23.75
1020	12" x 3 1/2"	"	29.25
1040	9" x 5 1/2"	"	35.25
1050	12" x 5 1/2"	"	42.75
1070	18" x 5 1/2"	"	65.75
2000	Add for Architectural Grade	PCT.	26.05
2010	Add for 6 3/4"	"	41.68
3000	Micro Lam - Plywood - 24" O.C.		
3010	9 1/2" x 1 3/4"	S.F.	6.38
3020	11 7/8" x 1 3/4"	"	6.85
3040	16" x 1 3/4"	"	8.91
4000	Glue Lam - Dimension Lumber - 24" O.C.		
4010	9" x 3 1/2"	S.F.	11.81
4020	12" x 3 1/2"	"	14.62
4040	9" x 5 1/2"	"	17.56
4050	12" x 5 1/2"	"	21.48
4070	18" x 5 1/2"	"	32.50
06999.60	**WOOD TREATMENTS**		
0010	PRESERVATIVES – PRESSURE TREATED		
0020	Dimensions	B.F.	0.33
0030	Timbers	"	0.49
0040	FIRE RETARDENTS		
0050	Dimensions & Timbers	B.F.	0.63
06999.70	**ROUGH HARDWARE**		
1000	NAILS	B.F.	0.03
2000	JOIST HANGERS	EA.	4.12
3000	BOLTS - 5/8" x 12"	"	5.38

DCD

Design Cost Data™

TABLE OF CONTENTS   PAGE

		UNIT	LABOR	MAT.	TOTAL
07100.10	**WATERPROOFING**				
0100	Membrane waterproofing, elastomeric				
1020	Butyl				
1040	1/32" thick	S.F.	1.97	1.47	3.44
1060	1/16" thick	"	2.05	1.91	3.96
1140	Neoprene				
1160	1/32" thick	S.F.	1.97	2.51	4.48
1180	1/16" thick	"	2.05	3.60	5.65
1260	Plastic vapor barrier (polyethylene)				
1280	4 mil	S.F.	0.19	0.05	0.24
1300	6 mil	"	0.19	0.09	0.28
1320	10 mil	"	0.24	0.12	0.36
1400	Bituminous membrane, asphalt felt, 15 lb.				
1440	One ply	S.F.	1.23	0.87	2.10
1460	Two ply	"	1.49	1.03	2.52
1480	Three ply	"	1.76	1.27	3.03
07160.10	**BITUMINOUS DAMPPROOFING**				
0100	Building paper, asphalt felt				
0120	15 lb	S.F.	1.97	0.19	2.16
0140	30 lb	"	2.05	0.37	2.42
1000	Asphalt, troweled, cold, primer plus				
1020	1 coat	S.F.	1.64	0.68	2.32
1040	2 coats	"	2.46	1.43	3.89
1060	3 coats	"	3.08	2.04	5.12
1200	Fibrous asphalt, hot troweled, primer plus				
1220	1 coat	S.F.	1.97	0.68	2.65
1240	2 coats	"	2.74	1.43	4.17
1260	3 coats	"	3.52	2.04	5.56
07190.10	**VAPOR BARRIERS**				
0980	Vapor barrier, polyethylene				
1000	2 mil	S.F.	0.24	0.01	0.25
1010	6 mil	"	0.24	0.06	0.30
1020	8 mil	"	0.27	0.07	0.34
1040	10 mil	"	0.27	0.08	0.35
07210.10	**BATT INSULATION**				
0980	Ceiling, fiberglass, unfaced				
1000	3-1/2" thick, R11	S.F.	0.58	0.42	1.00
1020	6" thick, R19	"	0.65	0.56	1.21
1030	9" thick, R30	"	0.75	1.10	1.85
1035	Suspended ceiling, unfaced				
1040	3-1/2" thick, R11	S.F.	0.54	0.42	0.96
1060	6" thick, R19	"	0.61	0.56	1.17
1070	9" thick, R30	"	0.70	1.10	1.80
1075	Crawl space, unfaced				
1080	3-1/2" thick, R11	S.F.	0.75	0.42	1.17
1100	6" thick, R19	"	0.82	0.56	1.38
1120	9" thick, R30	"	0.89	1.10	1.99
2000	Wall, fiberglass				
2010	Paper backed				
2020	2" thick, R7	S.F.	0.51	0.27	0.78
2040	3" thick, R8	"	0.54	0.30	0.84
2060	4" thick, R11	"	0.58	0.49	1.07

		UNIT	LABOR	MAT.	TOTAL
07210.10	**BATT INSULATION, Cont'd...**				
2080	6" thick, R19	S.F.	0.61	0.73	1.34
2090	Foil backed, 1 side				
2100	2" thick, R7	S.F.	0.51	0.63	1.14
2120	3" thick, R11	"	0.54	0.68	1.22
2140	4" thick, R14	"	0.58	0.71	1.29
2160	6" thick, R21	"	0.61	0.93	1.54
07210.20	**BOARD INSULATION**				
2200	Perlite board, roof				
2220	1.00" thick, R2.78	S.F.	0.41	0.63	1.04
2240	1.50" thick, R4.17	"	0.42	0.99	1.41
2580	Rigid urethane				
2600	1" thick, R6.67	S.F.	0.41	1.20	1.61
2640	1.50" thick, R11.11	"	0.42	1.63	2.05
2780	Polystyrene				
2800	1.0" thick, R4.17	S.F.	0.41	0.45	0.86
2820	1.5" thick, R6.26	"	0.42	0.70	1.12
07210.60	**LOOSE FILL INSULATION**				
1000	Blown-in type				
1010	Fiberglass				
1020	5" thick, R11	S.F.	0.41	0.41	0.82
1040	6" thick, R13	"	0.49	0.48	0.97
1060	9" thick, R19	"	0.70	0.58	1.28
07210.70	**SPRAYED INSULATION**				
1000	Foam, sprayed on				
1010	Polystyrene				
1020	1" thick, R4	S.F.	0.49	0.69	1.18
1040	2" thick, R8	"	0.65	1.34	1.99
1050	Urethane				
1060	1" thick, R4	S.F.	0.49	0.65	1.14
1080	2" thick, R8	"	0.65	1.24	1.89
07310.10	**ASPHALT SHINGLES**				
1000	Standard asphalt shingles, strip shingles				
1020	210 lb/square	SQ.	60.00	90.00	150
1040	235 lb/square	"	67.00	95.00	162
1060	240 lb/square	"	75.00	99.00	174
1080	260 lb/square	"	86.00	140	226
1100	300 lb/square	"	100	150	250
1120	385 lb/square	"	120	210	330
5980	Roll roofing, mineral surface				
6000	90 lb	SQ.	43.00	55.00	98.00
6020	110 lb	"	50.00	92.00	142
6040	140 lb	"	60.00	95.00	155
07310.50	**METAL SHINGLES**				
0980	Aluminum, .020" thick				
1000	Plain	SQ.	120	240	360
1020	Colors	"	120	270	390
1960	Steel, galvanized				
1980	26 ga.				
2000	Plain	SQ.	120	300	420
2020	Colors	"	120	380	500
2030	24 ga.				

		UNIT	LABOR	MAT.	TOTAL
07310.50	**METAL SHINGLES, Cont'd...**				
2040	Plain	SQ.	120	350	470
2060	Colors	"	120	440	560
07310.60	**SLATE SHINGLES**				
0960	Slate shingles				
0980	Pennsylvania				
1000	Ribbon	SQ.	300	600	900
1020	Clear	"	300	770	1,070
1030	Vermont				
1040	Black	SQ.	300	710	1,010
1060	Gray	"	300	780	1,080
1070	Green	"	300	800	1,100
1080	Red	"	300	1,440	1,740
07310.70	**WOOD SHINGLES**				
1000	Wood shingles, on roofs				
1010	White cedar, #1 shingles				
1020	4" exposure	SQ.	200	240	440
1040	5" exposure	"	150	220	370
1050	#2 shingles				
1060	4" exposure	SQ.	200	170	370
1080	5" exposure	"	150	150	300
1090	Resquared and rebutted				
1100	4" exposure	SQ.	200	220	420
1120	5" exposure	"	150	180	330
1140	On walls				
1150	White cedar, #1 shingles				
1160	4" exposure	SQ.	300	240	540
1180	5" exposure	"	240	220	460
1200	6" exposure	"	200	180	380
1210	#2 shingles				
1220	4" exposure	SQ.	300	170	470
1240	5" exposure	"	240	150	390
1260	6" exposure	"	200	120	320
1300	Add for fire retarding	"			110
07310.80	**WOOD SHAKES**				
2010	Shakes, hand split, 24" red cedar, on roofs				
2020	5" exposure	SQ.	300	280	580
2040	7" exposure	"	240	260	500
2060	9" exposure	"	200	240	440
2080	On walls				
2100	6" exposure	SQ.	300	260	560
2120	8" exposure	"	240	250	490
2140	10" exposure	"	200	230	430
3000	Add for fire retarding	"			71.00
07310.90	**CLAY TILE ROOF (assorted colors)**				
1010	Spanish tile, terracota	SQ.	150	420	570
1020	Mission tile, terracota	"	200	820	1,020
1030	French tile, rustic blue, green patina	"	170	950	1,120
07460.10	**METAL SIDING PANELS**				
1000	Aluminum siding panels				
1020	Corrugated				
1030	Plain finish				

		UNIT	LABOR	MAT.	TOTAL
07460.10	**METAL SIDING PANELS, Cont'd...**				
1040	.024"	S.F.	2.78	2.23	5.01
1060	.032"	"	2.78	2.62	5.40
1070	Painted finish				
1080	.024"	S.F.	2.78	2.78	5.56
1100	.032"	"	2.78	3.19	5.97
2000	Steel siding panels				
2040	Corrugated				
2080	22 ga.	S.F.	4.64	2.48	7.12
2100	24 ga.	"	4.64	2.26	6.90
07460.50	**PLASTIC SIDING**				
1000	Horizontal vinyl siding, solid				
1010	8" wide				
1020	Standard	S.F.	2.42	1.23	3.65
1040	Insulated	"	2.42	1.49	3.91
1050	10" wide				
1060	Standard	S.F.	2.25	1.27	3.52
1080	Insulated	"	2.25	1.52	3.77
8500	Vinyl moldings for doors and windows	L.F.	2.52	0.79	3.31
07460.60	**PLYWOOD SIDING**				
1000	Rough sawn cedar, 3/8" thick	S.F.	2.10	1.96	4.06
1020	Fir, 3/8" thick	"	2.10	1.08	3.18
1980	Texture 1-11, 5/8" thick				
2000	Cedar	S.F.	2.25	2.66	4.91
2020	Fir	"	2.25	1.85	4.10
2040	Redwood	"	2.14	2.86	5.00
2060	Southern Yellow Pine	"	2.25	1.51	3.76
07460.80	**WOOD SIDING**				
1000	Beveled siding, cedar				
1010	A grade				
1040	1/2 x 8	S.F.	2.52	4.54	7.06
1060	3/4 x 10	"	2.10	5.84	7.94
1070	Clear				
1100	1/2 x 8	S.F.	2.52	5.05	7.57
1120	3/4 x 10	"	2.10	6.77	8.87
1130	B grade				
1160	1/2 x 8	S.F.	2.52	5.40	7.92
1180	3/4 x 10	"	2.10	5.09	7.19
2000	Board and batten				
2010	Cedar				
2020	1x6	S.F.	3.15	6.22	9.37
2040	1x8	"	2.52	5.66	8.18
2060	1x10	"	2.25	5.11	7.36
2080	1x12	"	2.03	4.58	6.61
2090	Pine				
2100	1x6	S.F.	3.15	1.57	4.72
2120	1x8	"	2.52	1.54	4.06
2140	1x10	"	2.25	1.47	3.72
2160	1x12	"	2.03	1.35	3.38
3000	Tongue and groove				
3010	Cedar				
3020	1x4	S.F.	3.50	5.84	9.34

		UNIT	LABOR	MAT.	TOTAL
07460.80	**WOOD SIDING, Cont'd...**				
3040	1x6	S.F.	3.31	5.62	8.93
3060	1x8	"	3.15	5.27	8.42
3080	1x10	"	3.00	5.17	8.17
3090	Pine				
3100	1x4	S.F.	3.50	1.75	5.25
3120	1x6	"	3.31	1.66	4.97
3140	1x8	"	3.15	1.55	4.70
3160	1x10	"	3.00	1.47	4.47
07510.10	**BUILT-UP ASPHALT ROOFING**				
0980	Built-up roofing, asphalt felt, including gravel				
1000	2 ply	SQ.	150	88.00	238
1500	3 ply	"	200	120	320
2000	4 ply	"	240	170	410
2195	Cant strip, 4" x 4"				
2200	Treated wood	L.F.	1.71	2.56	4.27
2260	Foamglass	"	1.50	2.20	3.70
8000	New gravel for built-up roofing, 400 lb/sq	SQ.	120	44.50	165
07530.10	**SINGLE-PLY ROOFING**				
2000	Elastic sheet roofing				
2060	Neoprene, 1/16" thick	S.F.	0.75	2.83	3.58
2115	PVC				
2120	45 mil	S.F.	0.75	2.03	2.78
2200	Flashing				
2220	Pipe flashing, 90 mil thick				
2260	1" pipe	EA.	15.00	34.00	49.00
2360	Neoprene flashing, 60 mil thick strip				
2380	6" wide	L.F.	5.01	1.72	6.73
2390	12" wide	"	7.52	3.38	10.90
07610.10	**METAL ROOFING**				
1000	Sheet metal roofing, copper, 16 oz, batten seam	SQ.	400	1,800	2,200
1020	Standing seam	"	380	1,760	2,140
2000	Aluminum roofing, natural finish				
2005	Corrugated, on steel frame				
2010	.0175" thick	SQ.	170	130	300
2040	.0215" thick	"	170	170	340
2060	.024" thick	"	170	200	370
2080	.032" thick	"	170	250	420
2100	V-beam, on steel frame				
2120	.032" thick	SQ.	170	260	430
2130	.040" thick	"	170	280	450
2140	.050" thick	"	170	350	520
2200	Ridge cap				
2220	.019" thick	L.F.	2.00	4.04	6.04
2500	Corrugated galvanized steel roofing, on steel frame				
2520	28 ga.	SQ.	170	220	390
2540	26 ga.	"	170	250	420
2550	24 ga.	"	170	290	460
2560	22 ga.	"	170	320	490

DIVISION # 07 THERMAL AND MOISTURE

		UNIT	LABOR	MAT.	TOTAL
07620.10	**FLASHING AND TRIM**				
0050	Counter flashing				
0060	Aluminum, .032"	S.F.	6.01	2.09	8.10
0100	Stainless steel, .015"	"	6.01	6.69	12.70
0105	Copper				
0110	16 oz.	S.F.	6.01	9.36	15.37
0112	20 oz.	"	6.01	11.00	17.01
0114	24 oz.	"	6.01	13.50	19.51
0116	32 oz.	"	6.01	16.50	22.51
0118	Valley flashing				
0120	Aluminum, .032"	S.F.	3.76	1.74	5.50
0130	Stainless steel, .015	"	3.76	5.56	9.32
0135	Copper				
0140	16 oz.	S.F.	3.76	9.36	13.12
0160	20 oz.	"	5.01	11.00	16.01
0180	24 oz.	"	3.76	13.50	17.26
0200	32 oz.	"	3.76	16.50	20.26
0380	Base flashing				
0400	Aluminum, .040"	S.F.	5.01	2.60	7.61
0410	Stainless steel, .018"	"	5.01	6.65	11.66
0415	Copper				
0420	16 oz.	S.F.	5.01	9.36	14.37
0422	20 oz.	"	3.76	11.00	14.76
0424	24 oz.	"	5.01	13.50	18.51
0426	32 oz.	"	5.01	16.50	21.51
07620.20	**GUTTERS AND DOWNSPOUTS**				
1500	Copper gutter and downspout				
1520	Downspouts, 16 oz. copper				
1530	Round				
1540	3" dia.	L.F.	4.01	12.50	16.51
1550	4" dia.	"	4.01	15.50	19.51
1800	Gutters, 16 oz. copper				
1810	Half round				
1820	4" wide	L.F.	6.01	11.25	17.26
1840	5" wide	"	6.68	13.75	20.43
1860	Type K				
1880	4" wide	L.F.	6.01	12.50	18.51
1890	5" wide	"	6.68	13.00	19.68
3000	Aluminum gutter and downspout				
3005	Downspouts				
3010	2" x 3"	L.F.	4.01	1.32	5.33
3030	3" x 4"	"	4.29	1.81	6.10
3035	4" x 5"	"	4.62	2.01	6.63
3038	Round				
3040	3" dia.	L.F.	4.01	2.22	6.23
3050	4" dia.	"	4.29	2.84	7.13
3240	Gutters, stock units				
3260	4" wide	L.F.	6.33	2.05	8.38
3270	5" wide	"	6.68	2.45	9.13
4101	Galvanized steel gutter and downspout				
4111	Downspouts, round corrugated				
4121	3" dia.	L.F.	4.01	1.90	5.91

		UNIT	LABOR	MAT.	TOTAL
07620.20	**GUTTERS AND DOWNSPOUTS, Cont'd...**				
4131	4" dia.	L.F.	4.01	2.55	6.56
4141	5" dia.	"	4.29	3.79	8.08
4151	6" dia.	"	4.29	5.03	9.32
4161	Rectangular				
4171	2" x 3"	L.F.	4.01	1.71	5.72
4191	3" x 4"	"	3.76	2.46	6.22
4201	4" x 4"	"	3.76	3.08	6.84
4300	Gutters, stock units				
4310	5" wide				
4320	Plain	L.F.	6.68	1.66	8.34
4330	Painted	"	6.68	1.80	8.48
4335	6" wide				
4340	Plain	L.F.	7.07	2.32	9.39
4360	Painted	"	7.07	2.59	9.66
07810.10	**PLASTIC SKYLIGHTS**				
1030	Single thickness, not including mounting curb				
1040	2' x 4'	EA.	75.00	410	485
1050	4' x 4'	"	100	550	650
1060	5' x 5'	"	150	730	880
1070	6' x 8'	"	200	1,560	1,760
07920.10	**CAULKING**				
0100	Caulk exterior, two component				
0120	1/4 x 1/2	L.F.	3.15	0.43	3.58
0140	3/8 x 1/2	"	3.50	0.66	4.16
0160	1/2 x 1/2	"	3.93	0.90	4.83
0220	Caulk interior, single component				
0240	1/4 x 1/2	L.F.	3.00	0.29	3.29
0260	3/8 x 1/2	"	3.31	0.41	3.72
0280	1/2 x 1/2	"	3.70	0.54	4.24

		UNIT	COST
07999.10	**WATERPROOFING**		
0010	1-Ply Membrane Felt 15#	S.F.	2.02
0020	2-Ply Membrane Felt 15#	"	3.07
0030	Hydrolithic	"	3.31
0040	Elastomeric Rubberized Asphalt with Poly Sheet	"	3.07
0060	Metallic Oxide 3-Coat	"	5.24
0070	Vinyl Plastic	"	3.67
0080	Bentonite 3/8" - Trowel	"	3.61
0090	5/8" - Panels	"	3.70
07999.20	**DAMPPROOFING**		
0010	Asphalt Trowel Mastic 1/16"	S.F.	1.23
0020	1/8"	"	1.66
0030	Spray Liquid 1-Coat	"	0.77
0040	2-Coat	"	1.08
0050	Brush Liquid 2-Coat	"	1.21
1000	Hot Mop 1-Coat and Primer	"	2.02
1010	1 Fibrous Asphalt	"	2.20
1020	Cementitious Per Coat 1/2" Coat	"	1.30
1030	Silicone 1-Coat	"	0.85
1040	2-Coat	"	1.49
1050	Add for Scaffold and Lift Operations	"	0.76
07999.30	**BUILDING INSULATION**		
1000	FLEXIBLE		
1010	Fiberglass 2 1/4" R 7.40	S.F.	0.95
1020	3 1/2" R 11.00	"	1.12
1040	6" R 19.00	"	1.41
1060	12" R 38.00	"	1.95
1100	Add for Ceiling Work	"	0.12
1110	Add for Paper Faced	"	0.14
1120	Add for Polystyrene Barrier (2m)	"	0.14
1130	Add for Scaffold Work	"	0.97
1200	RIGID		
1300	Fiberglass 1" R 4.35 3# Density	S.F.	1.48
1310	1 1/2" R 6.52 3# Density	"	1.90
1320	2" R 8.70 3# Density	"	2.25
1400	Styrene, Molded 1" R 4.3.5	"	6.81
1410	1 1/2" R 6.52	"	1.26
1420	2" R 7.69	"	1.69
1430	2" T&G R 7.69	"	2.17
1500	Styrene, Extruded 1" R 5.40	"	1.54
1510	1 1/2" R 6.52	"	1.82
1520	2" R 7.69	"	1.95
1530	2" T&G R 7.69	"	2.17
1600	Perlite 1" R 2.78	"	1.54
1610	2" R 5.56	"	2.46
1700	Urethane 1" R 6.67	"	1.82
1710	2" R 13.34	"	2.32
1720	Add for Glued Applications	"	0.12
1800	LOOSE 1" Styrene R 3.8	"	1.08
1810	1" Fiberglass R 2.2	"	1.04
1820	1" Rock Wool R 2.9	"	1.07
1830	1" Cellulose R 3.7	"	1.01

		UNIT	COST
07999.30	**BUILDING INSULATION, Cont'd...**		
1900	FOAMED Urethane Per Inch	S.F.	2.63
2000	SPRAYED Cellulose Per Inch	"	1.77
2100	Polystyrene	"	2.12
2200	Urethane	"	3.01
2300	ALUMINUM PAPER	"	0.81
2400	VINYL FACED FIBERGLASS	"	2.41
07999.80	**SHINGLE ROOFING**		
1000	ASPHALT SHINGLES 235# Seal Down	S.F.	1.70
1010	300# Laminated	"	1.95
1020	325# Fire Resistant	"	2.79
1030	Timberline	"	2.72
1040	340# Tab Lock	"	3.21
2000	ASPHALT ROLL 90#	"	0.93
3000	FIBERGLASS SHINGLES 215#	"	1.65
3010	250#	"	1.86
4000	RED CEDAR 16" x 5" to Weather #1 Grade	"	4.83
4010	#2 Grade	"	4.30
4020	#3 Grade	"	3.07
4030	24" x 10" to Weather - Hand Splits - 1/2" x 3/4" #	"	15.33
4040	3/4" x 1 1/4" #	"	15.46
4050	Add for Fire Retardant	"	1.23
5000	METAL Aluminum 020 mil	"	4.85
6000	Anodized 020 Mil	"	5.61
7000	Steel, Enameled Colored	"	7.22
8000	Galvanized Colored	"	6.24
9000	Add to Above for 15# Felt Underlayment		
9010	Add for Base Starter	S.F.	0.89
9020	Add for Ice & Water Starter	"	1.48
9030	Add for Boston Ridge	"	3.35
9040	Add for Removal & Haul Away	"	1.28
9050	Add for Pitches Over 5-12 each Pitch Increase	"	0.07
9060	Add for Chimneys, Skylights and Bay Windows	EA.	97.00
07999.90	**BUILT-UP ROOFING**		
1000	MEMBRANE		
1010	3-Ply Asphalt & Gravel - R 10.0	SQ.	830
1020	R 16.6	"	850
1030	4-Ply - R 10.0	"	860
1040	R 16.6	"	870
1050	5-Ply - R 10.0	"	890
1060	R 16.6	"	930
2000	Add for Sheet Rock over Steel Deck - 5/8"	"	177
2010	Add for Fiberglass Insulation	"	59.50
2020	Add for Pitch and Gravel	"	91.75
2030	Add for Sloped Roofs	"	64.25
2040	Add for Thermal Barrier	"	49.50
2050	Add for Upside Down Roofing System	"	177
3000	SINGLE-PLY - 60M Butylene Roofing & Gravel Ballast - R 10.0	"	600
3010	Mech. Fastened - R 10.0	"	590
3020	PVC & EPDM Roofing & Gravel Ballast - R 10.0	"	560
3030	Mech. Fastened - R 10.0	"	590
3040	Blocking and Cants Not Included		

TABLE OF CONTENTS   PAGE

		UNIT	LABOR	MAT.	TOTAL
08110.10	**METAL DOORS**				
1000	Flush hollow metal, std. duty, 20 ga., 1-3/8" thick				
1020	2-6 x 6-8	EA.	70.00	350	420
1080	3-0 x 6-8	"	70.00	420	490
1090	1-3/4" thick				
1100	2-6 x 6-8	EA.	70.00	410	480
1150	3-0 x 6-8	"	70.00	470	540
1200	2-6 x 7-0	"	70.00	450	520
1240	3-0 x 7-0	"	70.00	500	570
2110	Heavy duty, 20 ga., unrated, 1-3/4"				
2130	2-8 x 6-8	EA.	70.00	450	520
2135	3-0 x 6-8	"	70.00	490	560
2140	2-8 x 7-0	"	70.00	520	590
2150	3-0 x 7-0	"	70.00	500	570
2200	18 ga., 1-3/4", unrated door				
2210	2-0 x 7-0	EA.	70.00	480	550
2235	2-6 x 7-0	"	70.00	480	550
2260	3-0 x 7-0	"	70.00	540	610
2270	3-4 x 7-0	"	70.00	560	630
2310	2", unrated door				
2320	2-0 x 7-0	EA.	79.00	530	609
2340	2-6 x 7-0	"	79.00	530	609
2360	3-0 x 7-0	"	79.00	600	679
2370	3-4 x 7-0	"	79.00	610	689
2400	Galvanized metal door				
2410	3-0 x 7-0	EA.	79.00	620	699
2450	For lead lining in doors	"			1,120
2460	For sound attenuation	"			100
4280	Vision glass				
4300	8" x 8"	EA.	79.00	130	209
4320	8" x 48"	"	79.00	200	279
4340	Fixed metal louver	"	63.00	290	353
4350	For fire rating, add				
4370	3 hr door	EA.			490
4380	1-1/2 hr door	"			220
4400	3/4 hr door	"			110
4430	1' extra height, add to material, 20%				
4440	1'6" extra height, add to material, 60%				
4470	For dutch doors with shelf, add to material, 100%				
5000	Stainless steel, general application				
5010	3'x7'	EA.	700	2,150	2,850
5020	5'X7'	"	930	3,110	4,040
6000	Heavy impact, s-core, 18 ga., stainless				
6010	3'x7'	EA.	700	2,270	2,970
6020	5'X7'	"	930	3,780	4,710
08110.40	**METAL DOOR FRAMES**				
1000	Hollow metal, stock, 18 ga., 4-3/4" x 1-3/4"				
1020	2-0 x 7-0	EA.	79.00	160	239
1060	2-6 x 7-0	"	79.00	190	269
1100	3-0 x 7-0	"	79.00	190	269
1120	4-0 x 7-0	"	110	210	320
1140	5-0 x 7-0	"	110	220	330

		UNIT	LABOR	MAT.	TOTAL
08110.40	**METAL DOOR FRAMES, Cont'd...**				
1160	6-0 x 7-0	EA.	110	260	370
1500	16 ga., 6-3/4" x 1-3/4"				
1520	2-0 x 7-0	EA.	87.00	190	277
1535	2-6 x 7-0	"	87.00	180	267
1550	3-0 x 7-0	"	87.00	200	287
1560	4-0 x 7-0	"	120	230	350
1580	6-0 x 7-0	"	120	260	380
08210.10	**WOOD DOORS**				
0980	Solid core, 1-3/8" thick				
1000	Birch faced				
1020	2-4 x 7-0	EA.	79.00	180	259
1060	3-0 x 7-0	"	79.00	180	259
1070	3-4 x 7-0	"	79.00	370	449
1080	2-4 x 6-8	"	79.00	180	259
1090	2-6 x 6-8	"	79.00	180	259
1100	3-0 x 6-8	"	79.00	180	259
1120	Lauan faced				
1140	2-4 x 6-8	EA.	79.00	160	239
1180	3-0 x 6-8	"	79.00	180	259
1200	3-4 x 6-8	"	79.00	190	269
1300	Tempered hardboard faced				
1320	2-4 x 7-0	EA.	79.00	200	279
1360	3-0 x 7-0	"	79.00	240	319
1380	3-4 x 7-0	"	79.00	250	329
1420	Hollow core, 1-3/8" thick				
1440	Birch faced				
1460	2-4 x 7-0	EA.	79.00	160	239
1500	3-0 x 7-0	"	79.00	170	249
1520	3-4 x 7-0	"	79.00	180	259
1600	Lauan faced				
1620	2-4 x 6-8	EA.	79.00	69.00	148
1630	2-6 x 6-8	"	79.00	74.00	153
1660	3-0 x 6-8	"	79.00	97.00	176
1680	3-4 x 6-8	"	79.00	110	189
1740	Tempered hardboard faced				
1770	2-6 x 7-0	EA.	79.00	90.00	169
1800	3-0 x 7-0	"	79.00	110	189
1820	3-4 x 7-0	"	79.00	120	199
1900	Solid core, 1-3/4" thick				
1920	Birch faced				
1940	2-4 x 7-0	EA.	79.00	280	359
1950	2-6 x 7-0	"	79.00	280	359
1970	3-0 x 7-0	"	79.00	270	349
1980	3-4 x 7-0	"	79.00	280	359
2000	Lauan faced				
2020	2-4 x 7-0	EA.	79.00	190	269
2030	2-6 x 7-0	"	79.00	220	299
2060	3-4 x 7-0	"	79.00	240	319
2080	3-0 x 7-0	"	79.00	260	339
2140	Tempered hardboard faced				
2170	2-6 x 7-0	EA.	79.00	280	359

		UNIT	LABOR	MAT.	TOTAL
08210.10	**WOOD DOORS, Cont'd...**				
2190	3-0 x 7-0	EA.	79.00	320	399
2200	3-4 x 7-0	"	79.00	340	419
2250	Hollow core, 1-3/4" thick				
2270	Birch faced				
2295	2-6 x 7-0	EA.	79.00	190	269
2320	3-0 x 7-0	"	79.00	200	279
2340	3-4 x 7-0	"	79.00	220	299
2400	Lauan faced				
2430	2-6 x 6-8	EA.	79.00	130	209
2460	3-0 x 6-8	"	79.00	120	199
2480	3-4 x 6-8	"	79.00	120	199
2520	Tempered hardboard				
2550	2-6 x 7-0	EA.	79.00	110	189
2580	3-0 x 7-0	"	79.00	120	199
2600	3-4 x 7-0	"	79.00	130	209
2620	Add-on, louver	"	63.00	35.00	98.00
2640	Glass	"	63.00	110	173
2700	Exterior doors, 3-0 x 7-0 x 2-1/2", solid core				
2710	Carved				
2720	One face	EA.	160	1,460	1,620
2740	Two faces	"	160	2,020	2,180
3000	Closet doors, 1-3/4" thick				
3001	Bi-fold or bi-passing, includes frame and trim				
3020	Paneled				
3040	4-0 x 6-8	EA.	110	510	620
3060	6-0 x 6-8	"	110	580	690
3070	Louvered				
3080	4-0 x 6-8	EA.	110	350	460
3100	6-0 x 6-8	"	110	420	530
3130	Flush				
3140	4-0 x 6-8	EA.	110	260	370
3160	6-0 x 6-8	"	110	330	440
3170	Primed				
3180	4-0 x 6-8	EA.	110	280	390
3200	6-0 x 6-8	"	110	310	420
3210	French Door, Dual-Tempered, Clear-Glass, 6'-8'				
3220	24"x80"x1-3/4"	EA.	130	460	590
3230	with Low-E glass	"	130	630	760
3240	30"x80"x1-3/4"	"	130	530	660
3250	with Low-E glass	"	130	630	760
3260	42"x80"x1-3/4"	"	130	690	820
3270	with Low-E glass	"	130	830	960
3280	24"x96"x1-3/4"	"	130	550	680
3290	with Low-E glass	"	130	690	820
3310	32"x96"x1-3/4"	"	130	620	750
3320	with Low-E glass	"	130	760	890
3330	French door, 10-lite, 1-3/4' thick, 6'-8" high				
3340	24" wide	EA.	130	490	620
3350	30" wide	"	130	600	730
3360	36" wide	"	130	610	740
3370	96" high, 24" wide	"	130	620	750
3380	30" wide	"	130	750	880

		UNIT	LABOR	MAT.	TOTAL
08210.10	**WOOD DOORS, Cont'd...**				
3390	36" wide	EA.	130	970	1,100
3400	French door, 1-lite, 1-3/4' thick, 6'-8" high				
3410	48" wide	EA.	130	1,680	1,810
3420	56" wide	"	130	2,000	2,130
3430	60" wide	"	130	2,330	2,460
3440	72" wide	"	130	2,200	2,330
3500	Fiberglass, single door, 6'-8" high				
3510	2-4" wide	EA.	79.00	240	319
3520	2-6" wide	"	79.00	360	439
3530	2-8" wide	"	79.00	360	439
3540	3-0" wide	"	79.00	440	519
3550	8'-0" high				
3560	2-4" wide	EA.	79.00	480	559
3570	2-6" wide	"	79.00	610	689
3580	2-8" wide	"	79.00	730	809
3590	3-0" wide	"	79.00	690	769
3600	Fiberglass, double door, 6'-8" high				
3610	56" wide	EA.	130	850	980
3620	60" wide	"	130	970	1,100
3630	64" wide	"	130	1,000	1,130
3640	72" wide	"	130	1,150	1,280
3650	8'-0" high				
3700	56" wide	EA.	130	310	440
3710	60" wide	"	130	330	460
3720	64" wide	"	130	350	480
3730	72" wide	"	130	380	510
3740	Metal clad, double door, 6'-8" high				
3750	56" wide	EA.	130	320	450
3760	60" wide	"	130	340	470
3770	64" wide	"	130	340	470
3780	72" wide	"	130	360	490
3790	8'-0" high				
3800	56" wide	EA.	130	540	670
3810	60" wide	"	130	570	700
3820	64" wide	"	130	640	770
3830	72" wide	"	130	700	830
3840	For outswinging doors, add min.	"			80.00
3850	Maximum	"			170
08210.90	**WOOD FRAMES**				
0080	Frame, interior, pine				
0100	2-6 x 6-8	EA.	90.00	100	190
0160	3-0 x 6-8	"	90.00	120	210
0180	5-0 x 6-8	"	90.00	120	210
0200	6-0 x 6-8	"	90.00	130	220
0220	2-6 x 7-0	"	90.00	120	210
0260	3-0 x 7-0	"	90.00	140	230
0280	5-0 x 7-0	"	130	150	280
0300	6-0 x 7-0	"	130	160	290
1000	Exterior, custom, with threshold, including trim				
1040	Walnut				
1060	3-0 x 7-0	EA.	160	420	580

		UNIT	LABOR	MAT.	TOTAL
08210.90	**WOOD FRAMES, Cont'd...**				
1080	6-0 x 7-0	EA.	160	480	640
1090	Oak				
1100	3-0 x 7-0	EA.	160	380	540
1120	6-0 x 7-0	"	160	430	590
1200	Pine				
1240	2-6 x 7-0	EA.	130	160	290
1300	3-0 x 7-0	"	130	180	310
1320	3-4 x 7-0	"	130	200	330
1340	6-0 x 7-0	"	210	210	420
3000	Fire-rated wood jambs				
3010	20-Minute, positive or neutral pressure, 3/4"	L.F.	5.25	10.75	16.00
3020	45-60-minute	"	5.25	12.00	17.25
3030	90-minute	"	5.25	13.25	18.50
3040	120-minute	"	5.25	16.00	21.25
08300.10	**SPECIAL DOORS**				
1000	Vault door and frame, class 5, steel	EA.	630	7,560	8,190
1480	Overhead door, coiling insulated				
1500	Chain gear, no frame, 12' x 12'	EA.	790	3,410	4,200
2000	Aluminum, bronze glass panels, 12-9 x 13-0	"	630	3,850	4,480
2200	Garage, flush, ins. metal, primed, 9-0 x 7-0	"	210	1,040	1,250
3000	Sliding fire doors, motorized, fusible link, 3 hr.				
3040	3-0 x 6-8	EA.	1,260	5,620	6,880
3060	3-8 x 6-8	"	1,260	5,690	6,950
3080	4-0 x 8-0	"	1,260	5,790	7,050
3100	5-0 x 8-0	"	1,260	5,900	7,160
3200	Metal clad doors, including electric motor				
3210	Light duty				
3220	Minimum	S.F.	10.50	45.00	55.50
3240	Maximum	"	25.25	73.00	98.25
3250	Heavy duty				
3260	Minimum	S.F.	31.50	70.00	102
3280	Maximum	"	39.25	110	149
3500	Counter doors, (roll-up shutters), std, manual				
3510	Opening, 4' high				
3520	4' wide	EA.	530	1,300	1,830
3540	6' wide	"	530	1,760	2,290
3560	8' wide	"	570	1,980	2,550
3580	10' wide	"	790	2,200	2,990
3590	14' wide	"	790	2,750	3,540
3595	6' high				
3600	4' wide	EA.	530	1,540	2,070
3620	6' wide	"	570	2,010	2,580
3630	8' wide	"	630	2,200	2,830
3640	10' wide	"	790	2,480	3,270
3650	14' wide	"	900	2,810	3,710
3660	For stainless steel, add to material, 40%				
3670	For motor operator, add	EA.			1,520
3800	Service doors, (roll up shutters), std, manual				
3810	Opening				
3820	8' high x 8' wide	EA.	350	1,650	2,000
3840	12' high x 12' wide	"	790	2,310	3,100

		UNIT	LABOR	MAT.	TOTAL
08300.10	**SPECIAL DOORS, Cont'd...**				
3860	16' high x 14' wide	EA.	1,050	4,240	5,290
3870	20' high x 14' wide	"	1,580	4,790	6,370
3880	24' high x 16' wide	"	1,400	7,810	9,210
3890	For motor operator				
3900	Up to 12-0 x 12-0, add	EA.			1,550
3920	Over 12-0 x 12-0, add	"			1,980
4040	Roll-up doors				
4050	13-0 high x 14-0 wide	EA.	900	1,560	2,460
4060	12-0 high x 14-0 wide	"	900	1,980	2,880
5200	Sectional wood overhead, frames not incl.				
5220	Commercial grade, HD, 1-3/4" thick, manual				
5240	8' x 8'	EA.	530	1,080	1,610
5260	10' x 10'	"	570	1,560	2,130
5280	12' x 12'	"	630	2,030	2,660
5290	Chain hoist				
5300	12' x 16' high	EA.	1,050	3,000	4,050
5320	14' x 14' high	"	790	3,290	4,080
5340	20' x 8' high	"	1,260	2,830	4,090
5360	16' high	"	1,580	6,350	7,930
5990	Sectional metal overhead doors, complete				
6000	Residential grade, manual				
6020	9' x 7'	EA.	250	740	990
6040	16' x 7'	"	310	1,330	1,640
6100	Commercial grade				
6120	8' x 8'	EA.	530	860	1,390
6140	10' x 10'	"	570	1,150	1,720
6160	12' x 12'	"	630	1,910	2,540
6180	20' x 14', with chain hoist	"	1,260	4,560	5,820
6400	Sliding glass doors				
6420	Tempered plate glass, 1/4" thick				
6430	6' wide				
6440	Economy grade	EA.	210	1,180	1,390
6450	Premium grade	"	210	1,350	1,560
6455	12' wide				
6460	Economy grade	EA.	310	1,650	1,960
6465	Premium grade	"	310	2,480	2,790
6470	Insulating glass, 5/8" thick				
6475	6' wide				
6480	Economy grade	EA.	210	1,450	1,660
6500	Premium grade	"	210	1,860	2,070
6505	12' wide				
6510	Economy grade	EA.	310	1,800	2,110
6515	Premium grade	"	310	2,890	3,200
6520	1" thick				
6525	6' wide				
6530	Economy grade	EA.	210	1,820	2,030
6540	Premium grade	"	210	2,100	2,310
6545	12' wide				
6550	Economy grade	EA.	310	2,830	3,140
6560	Premium grade	"	310	4,140	4,450
6600	Added costs				
6620	Custom quality, add to material, 30%				

		UNIT	LABOR	MAT.	TOTAL
08300.10	**SPECIAL DOORS, Cont'd...**				
6630	Tempered glass, 6' wide, add	S.F.			5.08
6880	Residential storm door				
6900	Minimum	EA.	110	180	290
6920	Average	"	110	240	350
6940	Maximum	"	160	530	690
08410.10	**STOREFRONTS**				
0135	Storefront, aluminum and glass				
0140	Minimum	S.F.	8.70	29.25	37.95
0150	Average	"	9.94	43.75	53.69
0160	Maximum	"	11.50	87.00	98.50
1020	Entrance doors, premium, closers, panic dev.,etc.				
1030	1/2" thick glass				
1040	3' x 7'	EA.	580	3,890	4,470
1060	6' x 7'	"	870	6,650	7,520
1065	3/4" thick glass				
1070	3' x 7'	EA.	580	4,030	4,610
1080	6' x 7'	"	870	6,720	7,590
1085	1" thick glass				
1090	3' x 7'	EA.	580	4,370	4,950
1100	6' x 7'	"	870	7,720	8,590
1150	Revolving doors				
1151	7' diameter, 7' high				
1160	Minimum	EA.	6,810	27,660	34,470
1170	Average	"	10,900	34,800	45,700
1180	Maximum	"	13,620	44,910	58,530
08510.10	**STEEL WINDOWS**				
0100	Steel windows, primed				
1000	Casements				
1010	Operable				
1020	Minimum	S.F.	4.09	49.25	53.34
1040	Maximum	"	4.64	74.00	78.64
1060	Fixed sash	"	3.48	39.50	42.98
1080	Double hung	"	3.86	74.00	77.86
1100	Industrial windows				
1120	Horizontally pivoted sash	S.F.	4.64	63.00	67.64
1130	Fixed sash	"	3.86	49.25	53.11
1135	Security sash				
1140	Operable	S.F.	4.64	78.00	82.64
1150	Fixed	"	3.86	69.00	72.86
1155	Picture window	"	3.86	33.50	37.36
1160	Projecting sash				
1170	Minimum	S.F.	4.35	58.00	62.35
1180	Maximum	"	4.35	72.00	76.35
1930	Mullions	L.F.	3.48	15.25	18.73
08520.10	**ALUMINUM WINDOWS**				
0110	Jalousie				
0120	3-0 x 4-0	EA.	87.00	390	477
0140	3-0 x 5-0	"	87.00	450	537
0220	Fixed window				
0240	6 sf to 8 sf	S.F.	9.94	19.25	29.19
0250	12 sf to 16 sf	"	7.73	17.00	24.73

		UNIT	LABOR	MAT.	TOTAL
08520.10	**ALUMINUM WINDOWS, Cont'd...**				
0255	Projecting window				
0260	6 sf to 8 sf	S.F.	17.50	42.50	60.00
0270	12 sf to 16 sf	"	11.50	38.25	49.75
0275	Horizontal sliding				
0280	6 sf to 8 sf	S.F.	8.70	27.75	36.45
0290	12 sf to 16 sf	"	6.96	25.50	32.46
1140	Double hung				
1160	6 sf to 8 sf	S.F.	14.00	38.25	52.25
1180	10 sf to 12 sf	"	11.50	34.00	45.50
3010	Storm window, 0.5 cfm, up to				
3020	60 u.i. (united inches)	EA.	34.75	89.00	124
3060	80 u.i.	"	34.75	100	135
3080	90 u.i.	"	38.75	100	139
3100	100 u.i.	"	38.75	110	149
3110	2.0 cfm, up to				
3120	60 u.i.	EA.	34.75	110	145
3160	80 u.i.	"	34.75	120	155
3180	90 u.i.	"	38.75	130	169
3200	100 u.i.	"	38.75	130	169
08600.10	**WOOD WINDOWS**				
0980	Double hung				
0990	24" x 36"				
1000	Minimum	EA.	63.00	240	303
1002	Average	"	79.00	350	429
1004	Maximum	"	110	470	580
1010	24" x 48"				
1020	Minimum	EA.	63.00	280	343
1022	Average	"	79.00	410	489
1024	Maximum	"	110	570	680
1030	30" x 48"				
1040	Minimum	EA.	70.00	290	360
1042	Average	"	90.00	410	500
1044	Maximum	"	130	590	720
1050	30" x 60"				
1060	Minimum	EA.	70.00	320	390
1062	Average	"	90.00	510	600
1064	Maximum	"	130	630	760
1160	Casement				
1180	1 leaf, 22" x 38" high				
1220	Minimum	EA.	63.00	350	413
1222	Average	"	79.00	430	509
1224	Maximum	"	110	500	610
1230	2 leaf, 50" x 50" high				
1240	Minimum	EA.	79.00	940	1,019
1242	Average	"	110	1,230	1,340
1244	Maximum	"	160	1,410	1,570
1250	3 leaf, 71" x 62" high				
1260	Minimum	EA.	79.00	1,550	1,629
1262	Average	"	110	1,580	1,690
1264	Maximum	"	160	1,890	2,050
1290	5 leaf, 119" x 75" high				

		UNIT	LABOR	MAT.	TOTAL
08600.10	**WOOD WINDOWS, Cont'd...**				
1300	Minimum	EA.	90.00	2,670	2,760
1302	Average	"	130	2,880	3,010
1304	Maximum	"	210	3,680	3,890
1360	Picture window, fixed glass, 54" x 54" high				
1400	Minimum	EA.	79.00	550	629
1422	Average	"	90.00	620	710
1424	Maximum	"	110	1,100	1,210
1430	68" x 55" high				
1440	Minimum	EA.	79.00	990	1,069
1442	Average	"	90.00	1,140	1,230
1444	Maximum	"	110	1,490	1,600
1480	Sliding, 40" x 31" high				
1520	Minimum	EA.	63.00	330	393
1522	Average	"	79.00	500	579
1524	Maximum	"	110	600	710
1530	52" x 39" high				
1540	Minimum	EA.	79.00	410	489
1542	Average	"	90.00	610	700
1544	Maximum	"	110	650	760
1550	64" x 72" high				
1560	Minimum	EA.	79.00	630	709
1562	Average	"	110	1,010	1,120
1564	Maximum	"	130	1,110	1,240
1760	Awning windows				
1780	34" x 21" high				
1800	Minimum	EA.	63.00	330	393
1822	Average	"	79.00	380	459
1824	Maximum	"	110	440	550
1880	48" x 27" high				
1900	Minimum	EA.	70.00	410	480
1902	Average	"	90.00	490	580
1904	Maximum	"	130	570	700
1920	60" x 36" high				
1940	Minimum	EA.	79.00	430	509
1942	Average	"	110	760	870
1944	Maximum	"	130	860	990
8000	Window frame, milled				
8010	Minimum	L.F.	12.50	6.09	18.59
8020	Average	"	15.75	6.80	22.55
8030	Maximum	"	21.00	10.25	31.25
08710.10	**HINGES**				
1200	Hinges, material only				
1250	3 x 3 butts, steel, interior, plain bearing	PAIR			20.75
1260	4 x 4 butts, steel, standard	"			30.50
1270	5 x 4-1/2 butts, bronze/s. steel, heavy duty	"			79.00
1290	Pivot hinges				
1300	Top pivot	EA.			88.00
1310	Intermediate pivot	"			94.00
1320	Bottom pivot	"			180

		UNIT	LABOR	MAT.	TOTAL
08710.20	**LOCKSETS**				
1280	Latchset, heavy duty				
1300	Cylindrical	EA.	39.25	190	229
1320	Mortise	"	63.00	200	263
1325	Lockset, heavy duty				
1330	Cylindrical	EA.	39.25	310	349
1350	Mortise	"	63.00	350	413
2200	Preassembled locks and latches, brass				
2220	Latchset, passage or closet latch	EA.	53.00	280	333
2225	Lockset				
2230	Privacy (bath or bathroom)	EA.	53.00	340	393
2240	Entry lock	"	53.00	490	543
2285	Lockset				
2290	Privacy (bath or bedroom)	EA.	53.00	220	273
2300	Entry lock	"	53.00	240	293
08710.30	**CLOSERS**				
2600	Door closers				
2610	Standard	EA.	79.00	240	319
2620	Heavy duty	"	79.00	280	359
08710.40	**DOOR TRIM**				
1600	Panic device				
1610	Mortise	EA.	160	780	940
1620	Vertical rod	"	160	1,170	1,330
1630	Labeled, rim type	"	160	810	970
1640	Mortise	"	160	1,060	1,220
1650	Vertical rod	"	160	1,130	1,290
2300	Door plates				
2305	Kick plate, aluminum, 3 beveled edges				
2310	10" x 28"	EA.	31.50	29.00	60.50
2340	10" x 38"	"	31.50	37.75	69.25
2350	Push plate, 4" x 16"				
2360	Aluminum	EA.	12.50	26.50	39.00
2371	Bronze	"	12.50	84.00	96.50
2380	Stainless steel	"	12.50	67.00	79.50
2385	Armor plate, 40" x 34"	"	25.25	77.00	102
2388	Pull handle, 4" x 16"				
2390	Aluminum	EA.	12.50	94.00	107
2400	Bronze	"	12.50	180	193
2420	Stainless steel	"	12.50	140	153
2425	Hasp assembly				
2430	3"	EA.	10.50	4.40	14.90
2440	4-1/2"	"	14.00	5.50	19.50
2450	6"	"	18.00	8.74	26.74
08710.60	**WEATHERSTRIPPING**				
0100	Weatherstrip, head and jamb, metal strip, neoprene bulb				
0140	Standard duty	L.F.	3.50	4.95	8.45
0160	Heavy duty	"	3.93	5.50	9.43
3980	Spring type				
4000	Metal doors	EA.	160	55.00	215
4010	Wood doors	"	210	55.00	265
4020	Sponge type with adhesive backing	"	63.00	51.00	114
4500	Thresholds				

		UNIT	LABOR	MAT.	TOTAL
08710.60	**WEATHERSTRIPPING, Cont'd...**				
4510	Bronze	L.F.	15.75	53.00	68.75
4515	Aluminum				10.75
4520	Plain	L.F.	15.75	38.50	54.25
4525	Vinyl insert	"	15.75	39.25	55.00
4530	Aluminum with grit	"	15.75	37.50	53.25
4533	Steel				65.00
4535	Plain	L.F.	15.75	29.75	45.50
4540	Interlocking	"	53.00	39.50	92.50
08810.10	**GLAZING**				
0800	Sheet glass, 1/8" thick	S.F.	3.86	8.91	12.77
1020	Plate glass, bronze or grey, 1/4" thick	"	6.32	13.00	19.32
1040	Clear	"	6.32	10.25	16.57
1060	Polished	"	6.32	12.00	18.32
1980	Plexiglass				
2000	1/8" thick	S.F.	6.32	5.73	12.05
2020	1/4" thick	"	3.86	10.25	14.11
3000	Float glass, clear				
3010	3/16" thick	S.F.	5.80	6.93	12.73
3020	1/4" thick	"	6.32	7.07	13.39
3040	3/8" thick	"	8.70	14.25	22.95
3100	Tinted glass, polished plate, twin ground				
3120	3/16" thick	S.F.	5.80	9.55	15.35
3130	1/4" thick	"	6.32	9.55	15.87
3140	3/8" thick	"	8.70	15.25	23.95
5000	Insulated glass, bronze or gray				
5020	1/2" thick	S.F.	11.50	19.50	31.00
5040	1" thick	"	17.50	23.25	40.75
5100	Spandrel, polished, 1 side, 1/4" thick	"	6.32	15.50	21.82
6000	Tempered glass (safety)				
6010	Clear sheet glass				
6020	1/8" thick	S.F.	3.86	10.75	14.61
6030	3/16" thick	"	5.35	13.00	18.35
6040	Clear float glass				
6050	1/4" thick	S.F.	5.80	11.25	17.05
6060	5/16" thick	"	6.96	20.00	26.96
6070	3/8" thick	"	8.70	24.50	33.20
6080	1/2" thick	"	11.50	33.50	45.00
6800	Insulating glass, two lites, clear float glass				
6840	1/2" thick	S.F.	11.50	13.75	25.25
6850	5/8" thick	"	14.00	16.00	30.00
6860	3/4" thick	"	17.50	17.50	35.00
6870	7/8" thick	"	20.00	18.50	38.50
6880	1" thick	"	23.25	24.75	48.00
6885	Glass seal edge				
6890	3/8" thick	S.F.	11.50	11.75	23.25
6895	Tinted glass				
6900	1/2" thick	S.F.	11.50	23.75	35.25
6910	1" thick	"	23.25	25.50	48.75
6920	Tempered, clear				
6930	1" thick	S.F.	23.25	46.50	69.75
7200	Plate mirror glass				

		UNIT	LABOR	MAT.	TOTAL
08810.10	**GLAZING, Cont'd...**				
7205	1/4" thick				
7210	15 sf	S.F.	6.96	11.75	18.71
7220	Over 15 sf	"	6.32	10.75	17.07
9650	Sand-Blasted Glass				
9660	3/16" Float glass, full sandblast, no custom decoration	S.F.	5.80	12.25	18.05
9670	3/8" Float glass, full sandblast, no custom decoration	"	6.96	14.50	21.46
9680	Beveled glass				
9690	commercial standard grade	S.F.	6.96	150	157
08910.10	**GLAZED CURTAIN WALLS**				
1000	Curtain wall, aluminum system, framing sections				
1005	2" x 3"				
1010	Jamb	L.F.	5.80	17.75	23.55
1020	Horizontal	"	5.80	18.00	23.80
1030	Mullion	"	5.80	24.00	29.80
1035	2" x 4"				
1040	Jamb	L.F.	8.70	24.00	32.70
1060	Horizontal	"	8.70	24.75	33.45
1070	Mullion	"	8.70	24.00	32.70
1080	3" x 5-1/2"				
1090	Jamb	L.F.	8.70	31.75	40.45
1100	Horizontal	"	8.70	35.25	43.95
1110	Mullion	"	8.70	32.00	40.70
1115	4" corner mullion	"	11.50	42.50	54.00
1120	Coping sections				
1130	1/8" x 8"	L.F.	11.50	44.25	55.75
1140	1/8" x 9"	"	11.50	44.50	56.00
1150	1/8" x 12-1/2"	"	14.00	45.75	59.75
1160	Sill section				
1170	1/8" x 6"	L.F.	6.96	43.75	50.71
1180	1/8" x 7"	"	6.96	44.25	51.21
1190	1/8" x 8-1/2"	"	6.96	45.00	51.96
1200	Column covers, aluminum				
1210	1/8" x 26"	L.F.	17.50	43.75	61.25
1220	1/8" x 34"	"	18.25	44.25	62.50
1230	1/8" x 38"	"	18.25	44.50	62.75
1500	Doors				
1600	Aluminum framed, standard hardware				
1620	Narrow stile				
1630	2-6 x 7-0	EA.	350	810	1,160
1640	3-0 x 7-0	"	350	810	1,160
1660	3-6 x 7-0	"	350	840	1,190
1700	Wide stile				
1720	2-6 x 7-0	EA.	350	1,380	1,730
1730	3-0 x 7-0	"	350	1,490	1,840
1750	3-6 x 7-0	"	350	1,600	1,950
1800	Flush panel doors, to match adjacent wall panels				
1810	2-6 x 7-0	EA.	430	1,170	1,600
1820	3-0 x 7-0	"	430	1,230	1,660
1840	3-6 x 7-0	"	430	1,270	1,700
2100	Wall panel, insulated				
2120	"U"=.08	S.F.	5.80	14.75	20.55

		UNIT	LABOR	MAT.	TOTAL
08910.10	**GLAZED CURTAIN WALLS, Cont'd...**				
2140	"U"=.10	S.F.	5.80	14.00	19.80
2160	"U"=.15	"	5.80	12.50	18.30
3000	Window wall system, complete				
3010	Minimum	S.F.	6.96	42.75	49.71
3030	Average	"	7.73	68.00	75.73
3050	Maximum	"	9.94	160	170
4860	Added costs				
4870	For bronze, add 20% to material				
4880	For stainless steel, add 50% to material				

DIVISION # 08 DOORS AND WINDOWS - QUICK ESTIMATING

		UNIT	COST
08999.10	**DOORS**		
0900	HOLLOW METAL		
1000	CUSTOM FRAMES (16 ga)		
1010	2'6" x 6'8" x 4 3/4"	EA.	370
1020	2'8" x 6'8" x 4 3/4"	"	400
1030	3'0" x 6'8" x 4 3/4"	"	420
1040	3'4" x 6'8" x 4 3/4"	"	450
2000	Add for A, B or C Label	"	47.50
2010	Add for Side Lights	"	218
2020	Add for Frames over 7'0" - Height	"	88.25
2030	Add for Frames over 6 3/4" - Width	"	76.75
3000	CUSTOM DOORS (1 3/8" or 1 3/4") (18 ga)		
3010	2'6" x 6'8" x 1 3/4"	EA.	550
3020	2'8" x 6'8" x 1 3/4"	"	570
3030	3'0" x 6'8" x 1 3/4"	"	590
3040	3'4" x 6'8" x 1 3/4"	"	600
4000	Add for 16 ga	"	85.25
4010	Add for B and C Label	"	38.25
4020	Add for Vision Panels or Lights	"	164
5000	STOCK FRAMES (16 ga)		
5010	2'6" x 6'8" or 7'0" x 4 3/4"	EA.	248
5020	2'8" x 6'8" or 7'0" x 4 3/4"	"	261
5030	3'0" x 6'8" or 7'0" x 4 3/4"	"	271
5040	3'4" x 6'8" or 7'0" x 4 3/4"	"	291
5050	Add for A, B or C Label	"	45.25
6000	STOCK DOORS (1 3/8" or 1 3/4" - 18 ga)		
6010	2'6" x 6'8" or 7'0"	EA.	580
6020	2'8" x 6'8" or 7'0"	"	600
6030	3'0" x 6'8" or 7'0"	"	600
6040	3'4" x 6'8" or 7'0"	"	660
6050	Add for B or C Label	"	35.50
08999.20	**WOOD DOORS (Incl. butts and locksets)**		
1000	FLUSH DOORS - No Label - Paint Grade		
2000	Birch - 7 Ply 2'4" x 6'8" x 1 3/8" -Hollow Core	EA.	228
2010	2'6" x 6'8" x 1 3/8"	"	238
2020	2'8" x 6'8" x 1 3/8"	"	251
2030	3'0" x 6'8" x 1 3/8"	"	251
2040	2'4" x 6'8" x 1 3/4"	"	251
2050	2'6" x 6'8" x 1 3/4"	"	251
2060	2'8" x 6'8" x 1 3/4"	"	261
2070	3'0" x 6'8" x 1 3/4"	"	271
2080	3'4" x 6'8" x 1 3/4"	"	271
2090	3'6" x 6'8" x 1 3/4"	"	271
3000	Birch - 7 Ply 2'4" x 6'8" x 1 3/8" -Solid Core	"	271
3010	2'6" x 6'8" x 1 3/8"	"	271
3020	2'8" x 6'8" x 1 3/8"	"	311
3030	3'0" x 6'8" x 1 3/8"	"	311
3040	2'4" x 6'8" x 1 3/4"	"	311
3050	2'6" x 6'8" x 1 3/4"	"	311
3060	2'8" x 6'8" x 1 3/4"	"	340
3070	3'0" x 6'8" x 1 3/4"	"	340
3080	3'4" x 6'8" x 1 3/4"	"	350

		UNIT	COST
08999.20	**WOOD DOORS (Incl. butts and locksets), Cont'd...**		
3090	3'6" x 6'8" x 1 3/4"E	EA.	410
4000	Add for Stain Grade	"	44.50
4010	Add for 5 Ply	"	96.00
4020	Add for Jamb & Trim - Solid - Knock Down	"	99.00
4030	Add for Jamb & Trim - Veneer - Knock Down	"	65.50
4040	Add for Red Oak - Rotary Cut	"	39.50
4050	Add for Red Oak - Plain Sliced	"	53.00
5000	Deduct for Lauan	"	41.75
5010	Add for Architectural Grade - 7 Ply	"	93.00
5020	Add for Vinyl Overlay	"	71.75
5030	Add for 7' - 0" Doors	"	31.50
5040	Add for Lite Cutouts with Metal Frame	"	104
5050	Add for Wood Louvres	"	164
5060	Add for Transom Panels & Side Panels	"	208
6000	LABELED - 1 3/4" - Paint Grade		
6010	2'6" x 6'8" - Birch - 20 Min.	EA.	370
6020	2'8" x 6'8"	"	350
6030	3'0" x 6'8"	"	360
6040	3'4" x 6'8"	"	410
6050	3'6" x 6'8"	"	470
7000	2'6" x 6'8" - Birch - 45 Min	"	390
7010	2'8" x 6'8"	"	410
7020	3'0" x 6'8"	"	420
7030	3'4" x 6'8"	"	440
7040	3'6" x 6'8"	"	460
8000	2'6" x 6'8" - Birch - 60 Min	"	510
8010	2'8" x 6'8"	"	550
8020	3'0" x 6'8"	"	560
8030	3'4" x 6'8"	"	590
8040	3'6" x 6'8"	"	460
9000	PANEL DOORS		
9010	Exterior - 1 3/4" -Pine		
9020	2'8" x 6'8"	EA.	800
9030	3'0" x 6'8"	"	770
9040	Interior - 1 3/8" -Pine		
9050	2'6" x 6'8"	EA.	410
9060	2'8" x 6'8"	"	450
9070	3'0" x 6'8"	"	480
9080	Add for Sidelights	"	360
9140	Exterior - 1 3/4"- Birch/ Oak		
9150	2'8" x 6'8"	EA.	840
9160	3'0" x 6'8"	"	830
9170	Interior - 1 3/8"		
9180	2'6" x 6'8"	EA.	570
9190	2'8" x 6'8"	"	590
9200	3'0" x 6'8"	"	630
9210	Add for Sidelights	"	560
9212	LOUVERED DOORS - 1 3/8" -Pine		
9213	2'0" x 6'8"	EA.	320
9214	2'6" x 6'8"	"	360
9215	2'8" x 6'8"	"	380
9216	3'0" x 6'8"	"	400

		UNIT	COST
08999.20	**WOOD DOORS (Incl. butts and locksets), Cont'd...**		
9220	Birch/Oak - 1 3/8"		
9230	2'0" x 6'8"	EA.	660
9240	2'6" x 6'8"	"	670
9250	2'8" x 6'8"	"	700
9260	3'0" x 6'8"	"	710
9270	BI-FOLD with hdwr. 1 3/8" - Flush		
9280	Birch/Oak, 2'0" x 6'8"	EA.	231
9290	2'4" x 6'8"	"	260
9300	2'6" x 6'8"	"	270
9320	Lauan, 2'0" x 6'8"	"	231
9330	2'4" x 6'8"	"	241
9340	2'6" x 6'8"	"	251
9350	Add for Prefinished	"	28.75
9360	CAFE DOORS - 1 1/8" - Pair		
9370	2'6" x 3'8"	EA.	460
9380	2'8" x 3'8"	"	480
9390	3'0" x 3'8"	"	490
9400	FRENCH DOORS - 1 3/8" -Pine		
9410	2'6" x 3'8"	EA.	610
9420	2'8" x 3'8"	"	630
9430	3'0" x 3'8"	"	660
9440	2'6" x 3'8" -Birch/Oak	"	790
9450	2'8" x 3'8"	"	800
9460	3'0" x 3'8"	"	810
9470	DUTCH DOORS		
9480	2'6" x 3'8"	EA.	920
9490	2'8" x 3'8"	"	1,010
9500	3'0" x 3'8"	"	1,280
9510	PREHUNG DOORS (INCL. TRIM)		
9520	Exterior, entrance - 1 3/4"		
9530	Panel - 2'8" x 6'8"	EA.	820
9540	3'0" x 6'8"	"	900
9550	3'0" x 7'0"	"	1,000
9560	Add for Insulation	"	281
9570	Add for Side Lights	"	341
9580	Interior - 1 3/8"		
9590	Flush, H.C. - 2'6" x 6'8" -Birch/Oak	EA.	450
9600	2'8" x 6'8"	"	460
9610	3'0" x 6'8"Each	"	480
9620	Flush, H.C. - 2'6" x 6'8" -Lauan	"	350
9630	2'8" x 6'8"	"	390
9640	3'0" x 6'8"Each	"	390
9650	Add for Int. Panel Door	"	320
08999.30	**SPECIAL DOORS**		
1000	BI-FOLDING - PREHUNG, 1 3/8"		
1020	Wood - 2 Door - 2'0" x 6'8" -Oak/flush	EA.	370
1030	2'6" x 6'8"	"	410
1040	3'0" x 6'8"	"	430
1050	4 Door - 4'0" x 6'8"	"	470
1060	5'0" x 6'8"	"	480
1070	6'0" x 6'8"	"	560

		UNIT	COST
08999.30	**SPECIAL DOORS, Cont'd...**		
1080	Add for Prefinished	EA.	43.50
1090	Add for Jambs and Casings	"	107
2000	Wood - 2 Door - 2'0" x 6'8" Pine/Panel	"	710
2010	2'6" x 6'8"	"	730
2020	3'0" x 6'8"	"	840
2030	4 Door - 4'0" x 6'8"	"	1,050
2040	5'0" x 6'8"	"	1,150
2050	6'0" x 6'8"	"	1,400
3000	Metal - 2 Door - 2'0" x 6'8"	"	218
3010	2'6" x 6'8"	"	228
3020	3'0" x 6'8"	"	281
3030	4 Door - 4'0" x 6'8"	"	301
3040	5'0" x 6'8"	"	330
3050	6'0" x 6'8"	"	360
3060	Add for Plastic Overlay	"	48.50
3070	Add for Louvre or Decorative Type	"	250
4000	Leaded Mirror (Based on 2-Panel Unit)		
4010	4'0" x 6'8"	EA.	1,020
4020	5'0" x 6'8"	"	1,200
4030	6'0" x 6'8"	"	1,200
7000	ROLLING - DOORS & GRILLES		
7010	Doors - 8' x 8'	EA.	4,010
7020	10' x 10'	"	4,190
7030	Grilles - 6'-8" x 3'-2"	"	1,390
7040	6'-8" x 4'-2"	"	1,570
8000	SHOWER DOORS - 28" x 66"	"	420
9000	SLIDING OR PATIO DOORS		
9010	Metal (Aluminum) - Including Glass Thresholds & Screen		
9020	Opening 8'0" x 6'8"	EA.	2,370
9030	8'0" x 8'0"	"	2,690
9040	Wood		
9050	Vinyl Clad 6'0" x 6'10"	EA.	2,960
9060	8'0" x 6'10"	"	3,240
9070	Pine - Prefinished 6'0" x 6'10"	"	2,940
9080	8'0" x 6'10"	"	3,260
9090	Add for Grilles	"	480
9100	Add for Triple Glazing	"	311
9110	Add for Screen	"	218
9120	SOUND REDUCTION - Metal	"	2,850
9130	Wood	"	1,870
9200	VAULT DOORS - 6'6" x 2'2" with Frame - 2 hour	"	5,770
08999.50	**METAL WINDOWS**		
1000	EACH, 2'4" x 4'6" - ALUMINUM WINDOWS		
1010	Casement and Awning	EA.	528
1020	Sliding or Horizontal	"	438
1030	Double and Single-Hung	"	468
1040	or Vertical Sliding		
1050	Projected	EA.	498
1060	Add for Screens	"	74.75
1070	Add for Storms	"	131
1080	Add for Insulated Glass	"	114

		UNIT	COST
08999.50	**METAL WINDOWS, Cont'd...**		
2000	ALUMINUM SASH		
2010	Casement	EA.	458
2020	Sliding	"	408
2030	Single-Hung	"	408
2040	Projected	"	334
2050	Fixed	"	344
3000	STEEL WINDOWS		
3010	Double-Hung	EA.	890
3020	Projected	"	850
4000	STEEL SASH		
4010	Casement	EA.	651
4020	Double-Hung	"	721
4030	Projected	"	558
4040	Fixed	"	568
5000	By S.F., 2'4" x 4'6" - ALUMINUM WINDOWS		
5010	Casement and Awning	S.F.	50.78
5020	Sliding or Horizontal	"	40.05
5030	Double and Single-Hung	"	44.80
5040	or Vertical Sliding	"	36.25
5050	Projected	"	50.56
5060	Add for Screens	"	7.18
5070	Add for Storms	"	13.23
5080	Add for Insulated Glass	"	10.91
6000	ALUMINUM SASH		
6010	Casement	S.F.	47.41
6020	Sliding	"	40.05
6030	Single-Hung	"	40.05
6040	Projected	"	34.13
6050	Fixed	"	35.34
7000	STEEL WINDOWS		
7010	Double-Hung	S.F.	79.00
7020	Projected	"	75.50
8000	STEEL SASH		
8010	Casement	S.F.	61.51
8020	Double-Hung	"	75.50
8030	Projected	"	52.26
8040	Fixed	"	54.02
08999.60	**WOOD WINDOWS**		
1000	BASEMENT OR UTILITY		
1020	Prefinished with Screen, 2'8" x 1'4"	EA.	274
1030	2'8" x 2'0"	"	328
1040	Add for Double Glazing or Storm	"	70.25
2000	CASEMENT OR AWNING		
2010	Operating Units - Insulating Glass		
2020	Single 2'4" x 4'0"	EA.	780
2030	2'4" x 5'0"	"	740
3000	Double with Screen 4'0" x 4'0" (2 units)	"	990
3010	4'0" x 5'0" (2 units)	"	1,270
3020	Triple with Screen 6'0" x 4'0" (3 units)	"	1,710
3030	6'0" x 5'0" (3 units)	"	1,730
4000	Fixed Units - Insulating Glass		

		UNIT	COST
08999.60	**WOOD WINDOWS, Cont'd...**		
4010	Single 2'4" x 4'0"	EA.	640
4020	2'4" x 5'0"	"	720
4030	Picture 4'0" x 4'6"	"	880
4040	4'0" x 6'0"	"	1,040
5000	DOUBLE HUNG - Insul. Glass - 2'6" x 3'6"	"	690
5010	2'6" x 4'2"	"	690
5020	3'2" x 3'6"	"	710
5030	3'2" x 4'2"	"	780
5040	Add for Triple Glazing	"	109
6000	GLIDER - & Insul. Glass with Screen - 4'0" x 3'6"	"	1,590
6010	5'0" x 4'0"	"	1,770
8000	PICTURE WINDOWS		
8010	9'6" x 4'10"	EA.	3,330
8020	9'6" x 5'6"	"	2,550
9000	CASEMENT - Insulating Glass		
9010	30° - 5'10" x 4'2" - (3 units)	EA.	2,420
9015	7'10" x 4'2" - (4 units)	"	2,920
9020	5'10" x 5'2" - (3 units)	"	2,530
9030	7'10" x 5'2" - (4 units)	"	2,860
9040	45° - 5'4" x 4'2" - (3 units)	"	2,410
9050	7'4" x 4'2" - (4 units)	"	2,640
9060	5'4" x 5'2" - (3 units)	"	2,800
9070	7'4" x 5'2" - (4 units)	"	3,150
9100	CASEMENT BOW WINDOWS - Insulating Glass		
9110	6'2" x 4'2" - (3 units)	EA.	2,520
9120	8'2" x 4'2" - (4 units)	"	3,330
9130	6'2" x 5'2" - (3 units)	"	2,810
9140	8'2" x 5'2" - (4 units)	"	3,660
9150	Add for Triple Glazing - per unit	"	94.75
9160	Add for Bronze Glazing - per unit	"	94.75
9170	Deduct for Primed Only - per unit	"	23.82
9180	Add for Screens - per unit	"	29.17
9200	90° BOX BAY WINDOWS - Insulating Glass 4'8" x 4'2"	"	3,330
9210	6'8" x 4'2"	"	4,220
9220	6'8" x 5'2"	"	4,380
9300	ROOF WINDOWS 1'10" x 3'10"-Fixed	"	970
9310	2'4" x 3'10"	"	1,110
9320	3'8" x 3'10"	"	1,370
9400	ROOF WINDOWS 1'10" x 3'10"-Movable	"	1,300
9410	2'4" x 3'10"	"	1,600
9420	3'8" x 3'10"	"	1,840
9500	CIRCULAR TOPS & ROUNDS - 4'0"	"	1,040
9510	6'0"	"	2,320
08999.70	**SPECIAL WINDOWS**		
1000	LIGHT-PROOF WINDOWS	S.F.	59.75
2000	PASS WINDOWS	"	45.97
3000	DETENTION WINDOWS	"	62.25
4000	VENETIAN BLIND WINDOWS (ALUMINUM)	"	52.25
5000	SOUND-CONTROL WINDOWS	"	51.00

			UNIT	COST
08999.80	**DOOR AND WINDOW ACCESSORIES**			
1000	STORMS AND SCREENS			
1010	Windows			
1020	Screen Only - Wood - 3' x 5'		EA.	154
1030	Aluminum - 3' x 5'		"	179
1040	Storm & Screen Combination - Aluminum		"	202
1050	Doors		"	
1060	Screen Only - Wood		"	368
1070	Aluminum - 3' x 6' - 8'		"	368
1080	Storm & Screen Combination - Aluminum		"	441
1090	Wood 1 1/8"		"	490
2000	DETENTION SCREENS			
2010	Example: 4' 0" x 7' 0"		EA.	1,090
3000	DOOR OPENING ASSEMBLIES			
3010	Floor or Overhead Electric Eye Units			
3020	Swing - Single 3' x 7' Door - Hydraulic		EA.	6,560
3030	Double 6' x 7' Door - Hydraulic		"	10,430
3040	Sliding - Single 3' x 7' Door - Hydraulic		"	8,010
3050	Double 5' x 7' Doors- Hydraulic		"	10,380
3060	Industrial Doors - 10' x 8'		"	11,670
4000	SHUTTERS			
4010	16" x 1 1/8" x 48"		EA.	176
4020	16" x 1 1/8" x 60"		"	212
4030	16" x 1 1/8" x 72"		"	252
08999.90	**FINISH HARDWARE**			
1000	BUTT HINGES -Painted			
1010	3" x 3"		EA.	37.00
1020	3 1/2" x 3 1/2"		"	38.75
1030	4" x 4"		"	41.50
1040	4 1/2" x 4 1/2"		"	45.25
1050	4" x 4" Ball Bearing		"	68.00
1060	4 1/2" x 4 1/2" Ball Bearing		"	73.00
2000	Bronze			
2010	3" x 3"		EA.	38.75
2020	3 1/2" x 3 1/2"		"	39.75
2030	4" x 4"		"	41.50
2040	4 1/2" x 4 1/2"		"	48.00
2050	4" x 4" Ball Bearing		"	70.00
2060	4 1/2" x 4 1/2" Ball Bearing		"	79.00
3000	Chrome			
3010	3" x 3"		EA.	41.50
3020	3 1/2" x 3 1/2"		"	44.25
3030	4" x 4"		"	46.25
3040	4 1/2" x 4 1/2"		"	55.00
3050	4" x 4" Ball Bearing		"	79.00
3060	4 1/2" x 4 1/2" Ball Bearing		"	84.00
5000	CLOSERS - SURFACE MOUNTED - 3' - 0" Door		"	290
5010	3' - 4"		"	310
5020	3' - 8"		"	310
5030	4' - 0"		"	390
6000	CLOSERS - CONCEALED - Interior		"	520
6010	ExteriorEach		"	700

		UNIT	COST
08999.90	**FINISH HARDWARE, Cont'd...**		
6020	Add for Fusible Link - Electric	EA.	290
6030	CLOSERS - FLOOR HINGES - Interior	"	720
6040	Exterior	"	1,110
6050	Add for Hold Open Feature	"	115
6060	Add for Double Acting Feature	"	380
7000	DEAD BOLT LOCK - Cylinder - Outside Key	"	250
7010	Cylinder - Double Key	"	360
7020	Flush - Push/ Pull	"	95.00
8000	EXIT DEVICES (PANIC) - Surface	"	1,030
8010	Mortise Lock	"	1,340
8020	Concealed	"	3,330
8030	Handicap (ADA) Automatic	"	2,440
9000	HINGES, SPRING (PAINTED) - 6" Single Acting	"	181
9010	6" Double Acting	"	218
9101	LATCHSETS -Bronze or Chrome	"	290
9102	Stainless Steel	"	330
9200	LOCKSETS - Mortise - Bronze or Chrome - H.D.	"	400
9210	Mortise - Bronze or Chrome - S.D.	"	320
9220	Stainless Steel	"	510
9230	Cylindrical - Bronze or Chrome	"	350
9240	Stainless Steel	"	410
9300	LEVER HANDICAP - Latch Set	"	410
9310	Lock Set	"	470
9400	PLATES - Kick - 8" x 34" - Aluminum	"	127
9410	Bronze	"	117
9420	Push - 6" x 15" - Aluminum	"	82.25
9430	Bronze	"	119
9440	Push & Pull Combination - Aluminum	"	138
9450	Bronze	"	198
9500	STOPS AND HOLDERS		
9510	Holder - Magnetic (No Electric)	EA.	280
9520	Bumper	"	66.00
9530	Overhead - Bronze, Chrome or Aluminum	"	152
9540	Wall Stops	"	57.75
9550	Floor Stops	"	72.25
08999.91	**WEATHERSTRIPPING**		
1000	ASTRAGALS - Aluminum - 1/8" x 2"	EA.	77.75
1010	Painted Steel	"	70.50
2000	DOORS (WOOD) - Interlocking	"	125
2010	Spring Bronze	"	135
2020	Add for Metal Doors	"	62.25
3000	SWEEPS - 36" WOOD DOORS - Aluminum	"	36.25
3010	Vinyl	"	35.50
4000	THRESHOLDS - Aluminum - 4" x 1/2"	"	74.00
4010	Bronze 4" x 1/2"	"	94.25
4020	5 1/2" x 1/2"	"	100
5000	WINDOWS (WOOD) - Interlocking	"	107
5010	Spring Bronze	"	125

TABLE OF CONTENTS

09110.10	METAL STUDS	UNIT	LABOR	MAT.	TOTAL
0060	Studs, non load bearing, galvanized				
0130	3-5/8", 20 ga.				
0140	12" o.c.	S.F.	1.57	0.81	2.38
0142	16" o.c.	"	1.26	0.62	1.88
0144	24" o.c.	"	1.05	0.47	1.52
0170	25 ga.				
0180	12" o.c.	S.F.	1.57	0.53	2.10
0182	16" o.c.	"	1.26	0.44	1.70
0184	24" o.c.	"	1.05	0.33	1.38
0210	6", 20 ga.				
0220	12" o.c.	S.F.	1.96	1.14	3.10
0222	16" o.c.	"	1.57	0.83	2.40
0224	24" o.c.	"	1.31	0.68	1.99
0230	25 ga.				
0240	12" o.c.	S.F.	1.96	0.73	2.69
0242	16" o.c.	"	1.57	0.58	2.15
0244	24" o.c.	"	1.31	0.44	1.75
0980	Load bearing studs, galvanized				
0990	3-5/8", 16 ga.				
1000	12" o.c.	S.F.	1.57	1.47	3.04
1020	16" o.c.	"	1.26	1.36	2.62
1110	18 ga.				
1130	12" o.c.	S.F.	1.05	1.15	2.20
1140	16" o.c.	"	1.26	1.05	2.31
1980	6", 16 ga.				
2000	12" o.c.	S.F.	1.96	1.98	3.94
2001	16" o.c.	"	1.57	1.78	3.35
3000	Furring				
3160	On beams and columns				
3170	7/8" channel	L.F.	4.20	0.52	4.72
3180	1-1/2" channel	"	4.84	0.62	5.46
4460	On ceilings				
4470	3/4" furring channels				
4480	12" o.c.	S.F.	2.62	0.37	2.99
4490	16" o.c.	"	2.52	0.29	2.81
4495	24" o.c.	"	2.25	0.20	2.45
4500	1-1/2" furring channels				
4520	12" o.c.	S.F.	2.86	0.62	3.48
4540	16" o.c.	"	2.62	0.47	3.09
4560	24" o.c.	"	2.42	0.31	2.73
5000	On walls				
5020	3/4" furring channels				
5050	12" o.c.	S.F.	2.10	0.37	2.47
5100	16" o.c.	"	1.96	0.29	2.25
5150	24" o.c.	"	1.85	0.20	2.05
5200	1-1/2" furring channels				
5210	12" o.c.	S.F.	2.25	0.62	2.87
5220	16" o.c.	"	2.10	0.47	2.57
5230	24" o.c.	"	1.96	0.31	2.27

		UNIT	LABOR	MAT.	TOTAL
09205.10	**GYPSUM LATH**				
1070	Gypsum lath, 1/2" thick				
1090	Clipped	S.Y.	3.50	5.76	9.26
1110	Nailed	"	3.93	5.76	9.69
09205.20	**METAL LATH**				
0960	Diamond expanded, galvanized				
0980	2.5 lb., on walls				
1010	Nailed	S.Y.	7.87	3.83	11.70
1030	Wired	"	9.00	3.83	12.83
1040	On ceilings				
1050	Nailed	S.Y.	9.00	3.83	12.83
1070	Wired	"	10.50	3.83	14.33
2230	Stucco lath				
2240	1.8 lb.	S.Y.	7.87	4.51	12.38
2300	3.6 lb.	"	7.87	5.06	12.93
2310	Paper backed				
2320	Minimum	S.Y.	6.30	3.50	9.80
2400	Maximum	"	9.00	5.65	14.65
09205.60	**PLASTER ACCESSORIES**				
0120	Expansion joint, 3/4", 26 ga., galv.	L.F.	1.57	1.48	3.05
2000	Plaster corner beads, 3/4", galvanized	"	1.80	0.41	2.21
2020	Casing bead, expanded flange, galvanized	"	1.57	0.56	2.13
2100	Expanded wing, 1-1/4" wide, galvanized	"	1.57	0.66	2.23
2500	Joint clips for lath	EA.	0.31	0.17	0.48
2580	Metal base, galvanized, 2-1/2" high	L.F.	2.10	0.75	2.85
2600	Stud clips for gypsum lath	EA.	0.31	0.17	0.48
2700	Tie wire galvanized, 18 ga., 25 lb. hank	"			47.00
8000	Sound deadening board, 1/4"	S.F.	1.05	0.31	1.36
09210.10	**PLASTER**				
0980	Gypsum plaster, trowel finish, 2 coats				
1000	Ceilings	S.Y.	18.25	4.30	22.55
1020	Walls	"	17.25	4.30	21.55
1030	3 coats				
1040	Ceilings	S.Y.	25.50	5.96	31.46
1060	Walls	"	22.50	5.96	28.46
7000	On columns, add to installation, 50%	"			
7020	Chases, fascia, and soffits, add to installation, 50%	"			
7040	Beams, add to installation, 50%	"			
09220.10	**PORTLAND CEMENT PLASTER**				
2980	Stucco, portland, gray, 3 coat, 1" thick				
3000	Sand finish	S.Y.	25.50	8.03	33.53
3020	Trowel finish	"	26.75	8.03	34.78
3030	White cement				
3040	Sand finish	S.Y.	26.75	9.17	35.92
3060	Trowel finish	"	29.25	9.17	38.42
3980	Scratch coat				
4000	For ceramic tile	S.Y.	5.87	2.91	8.78
4020	For quarry tile	"	5.87	2.91	8.78
5000	Portland cement plaster				
5020	2 coats, 1/2"	S.Y.	11.75	5.78	17.53
5040	3 coats, 7/8"	"	14.75	6.90	21.65

09250.10 GYPSUM BOARD

		UNIT	LABOR	MAT.	TOTAL
0220	1/2", clipped to				
0240	Metal furred ceiling	S.F.	0.70	0.42	1.12
0260	Columns and beams	"	1.57	0.38	1.95
0270	Walls	"	0.63	0.38	1.01
0280	Nailed or screwed to				
0290	Wood or metal framed ceiling	S.F.	0.63	0.38	1.01
0300	Columns and beams	"	1.40	0.38	1.78
0400	Walls	"	0.57	0.38	0.95
1000	5/8", clipped to				
1020	Metal furred ceiling	S.F.	0.78	0.42	1.20
1040	Columns and beams	"	1.75	0.42	2.17
1060	Walls	"	0.70	0.42	1.12
1070	Nailed or screwed to				
1080	Wood or metal framed ceiling	S.F.	0.78	0.42	1.20
1100	Columns and beams	"	1.75	0.42	2.17
1120	Walls	"	0.70	0.42	1.12
1122	Vinyl faced, clipped to metal studs				
1124	1/2"	S.F.	0.78	1.19	1.97
1126	5/8"	"	0.78	1.13	1.91
1130	Add for				
1140	Fire resistant	S.F.			0.12
1180	Water resistant	"			0.19
1200	Water and fire resistant	"			0.24
1220	Taping and finishing joints				
1222	Minimum	S.F.	0.42	0.04	0.46
1224	Average	"	0.52	0.07	0.59
1226	Maximum	"	0.63	0.10	0.73
5020	Casing bead				
5022	Minimum	L.F.	1.80	0.16	1.96
5024	Average	"	2.10	0.18	2.28
5026	Maximum	"	3.15	0.22	3.37
5040	Corner bead				
5042	Minimum	L.F.	1.80	0.18	1.98
5044	Average	"	2.10	0.22	2.32
5046	Maximum	"	3.15	0.27	3.42

09310.10 CERAMIC TILE

		UNIT	LABOR	MAT.	TOTAL
0980	Glazed wall tile, 4-1/4" x 4-1/4"				
1000	Minimum	S.F.	4.29	2.32	6.61
1020	Average	"	5.01	3.68	8.69
1040	Maximum	"	6.01	13.25	19.26
1042	6" x 6"				
1044	Minimum	S.F.	3.75	1.65	5.40
1046	Average	"	4.29	2.22	6.51
1048	Maximum	"	5.01	2.77	7.78
2960	Base, 4-1/4" high				
2980	Minimum	L.F.	7.51	4.47	11.98
3000	Average	"	7.51	5.20	12.71
3040	Maximum	"	7.51	6.87	14.38
3042	Glazed moldings and trim, 12" x 12"				
3044	Minimum	L.F.	6.01	2.46	8.47
3046	Average	"	6.01	3.75	9.76

		UNIT	LABOR	MAT.	TOTAL
09310.10	**CERAMIC TILE, Cont'd...**				
3048	Maximum	L.F.	6.01	5.04	11.05
6100	Unglazed floor tile				
6120	Portland cem., cushion edge, face mtd				
6140	1" x 1"	S.F.	5.46	8.87	14.33
6150	2" x 2"	"	5.01	9.38	14.39
6162	4" x 4"	"	5.01	8.73	13.74
6164	6" x 6"	"	4.29	3.12	7.41
6166	12" x 12"	"	3.75	2.75	6.50
6168	16" x 16"	"	3.34	2.38	5.72
6170	18" x 18"	"	3.00	2.31	5.31
6200	Adhesive bed, with white grout				
6220	1" x 1"	S.F.	5.46	7.38	12.84
6230	2" x 2"	"	5.01	7.81	12.82
6260	4" x 4"	"	5.01	7.81	12.82
6262	6" x 6"	"	4.29	2.60	6.89
6264	12" x 12"	"	3.75	2.28	6.03
6266	16" x 16"	"	3.34	1.98	5.32
6268	18" x 18"	"	3.00	1.92	4.92
6300	Organic adhesive bed, thin set, back mounted				
6320	1" x 1"	S.F.	5.46	7.38	12.84
6350	2" x 2"	"	5.01	8.59	13.60
6360	For group 2 colors, add to material, 10%				
6370	For group 3 colors, add to material, 20%				
6380	For abrasive surface, add to material, 25%				
8990	Ceramic accessories				
9000	Towel bar, 24" long				
9004	Average	EA.	30.00	22.25	52.25
9020	Soap dish				
9024	Average	EA.	50.00	11.50	61.50
09330.10	**QUARRY TILE**				
1060	Floor				
1080	4 x 4 x 1/2"	S.F.	8.01	6.57	14.58
1120	6 x 6 x 3/4"	"	7.51	7.99	15.50
1200	Wall, applied to 3/4" portland cement bed				
1220	4 x 4 x 1/2"	S.F.	12.00	5.84	17.84
1240	6 x 6 x 3/4"	"	10.00	6.53	16.53
1320	Cove base				
1330	5 x 6 x 1/2" straight top	L.F.	10.00	6.66	16.66
1340	6 x 6 x 3/4" round top	"	10.00	6.18	16.18
1360	Stair treads 6 x 6 x 3/4"	"	15.00	9.13	24.13
1380	Window sill 6 x 8 x 3/4"	"	12.00	8.33	20.33
1400	For abrasive surface, add to material, 25%				
09410.10	**TERRAZZO**				
1100	Floors on concrete, 1-3/4" thick, 5/8" topping				
1120	Gray cement	S.F.	8.38	5.67	14.05
1140	White cement	"	8.38	5.92	14.30
1200	Sand cushion, 3" thick, 5/8" top, 1/4"				
1220	Gray cement	S.F.	9.78	6.69	16.47
1240	White cement	"	9.78	6.96	16.74
1260	Monolithic terrazzo, 3-1/2" base slab, 5/8" topping	"	7.34	4.75	12.09
1280	Terrazzo wainscot, cast-in-place, 1/2" thick	"	14.75	5.77	20.52

		UNIT	LABOR	MAT.	TOTAL
09410.10	**TERRAZZO, Cont'd...**				
1300	Base, cast in place, terrazzo cove type, 6" high	L.F.	8.38	8.47	16.85
1320	Curb, cast in place, 6" wide x 6" high, polished top	"	29.25	9.35	38.60
1400	For venetian type terrazzo, add to material, 10%				
1420	For abrasive heavy duty terrazzo, add to material, 15%				
1480	Divider strips				
1500	Zinc	L.F.			1.43
1510	Brass	"			2.66
09510.10	**CEILINGS AND WALLS**				
1520	Acoustical panels, suspension system not included				
1540	Fiberglass panels				
1550	5/8" thick				
1560	2' x 2'	S.F.	0.90	1.61	2.51
1580	2' x 4'	"	0.70	1.34	2.04
1590	3/4" thick				
1600	2' x 2'	S.F.	0.90	2.14	3.04
1620	2' x 4'	"	0.70	2.07	2.77
1640	Glass cloth faced fiberglass panels				
1660	3/4" thick	S.F.	1.05	3.05	4.10
1680	1" thick	"	1.05	3.41	4.46
1700	Mineral fiber panels				
1710	5/8" thick				
1720	2' x 2'	S.F.	0.90	1.37	2.27
1740	2' x 4'	"	0.70	1.37	2.07
1750	3/4" thick				
1760	2' x 2'	S.F.	0.90	2.14	3.04
1780	2' x 4'	"	0.70	2.07	2.77
3000	Acoustical tiles, suspension system not included				
3020	Fiberglass tile, 12" x 12"				
3040	5/8" thick	S.F.	1.14	2.00	3.14
3060	3/4" thick	"	1.40	2.32	3.72
3080	Glass cloth faced fiberglass tile				
3100	3/4" thick	S.F.	1.40	3.73	5.13
3120	3" thick	"	1.57	4.17	5.74
3140	Mineral fiber tile, 12" x 12"				
3150	5/8" thick				
3160	Standard	S.F.	1.26	1.05	2.31
3180	Vinyl faced	"	1.26	2.08	3.34
3190	3/4" thick				
3195	Standard	S.F.	1.26	1.53	2.79
3200	Vinyl faced	"	1.26	2.66	3.92
5500	Ceiling suspension systems				
5505	T bar system				
5510	2' x 4'	S.F.	0.63	1.15	1.78
5520	2' x 2'	"	0.70	1.25	1.95
5530	Concealed Z bar suspension system, 12" module	"	1.05	1.18	2.23
5550	For 1-1/2" carrier channels, 4' o.c., add	"			0.38
5560	Carrier channel for recessed light fixtures	"			0.69
09550.10	**WOOD FLOORING**				
0100	Wood strip flooring, unfinished				
1000	Fir floor				
1010	C and better				

		UNIT	LABOR	MAT.	TOTAL
09550.10	**WOOD FLOORING, Cont'd...**				
1020	Vertical grain	S.F.	2.10	3.20	5.30
1040	Flat grain	"	2.10	3.02	5.12
1060	Oak floor				
1080	Minimum	S.F.	3.00	3.38	6.38
1100	Average	"	3.00	4.66	7.66
1120	Maximum	"	3.00	6.75	9.75
1340	Added costs				
1350	For factory finish, add to material, 10%				
1355	For random width floor, add to total, 20%				
1360	For simulated pegs, add to total, 10%				
3000	Gym floor, 2 ply felt, 25/32" maple, finished, in mastic	S.F.	3.50	8.54	12.04
3020	Over wood sleepers	"	3.93	8.69	12.62
9020	Finishing, sand, fill, finish, and wax	"	1.57	0.66	2.23
9100	Refinish sand, seal, and 2 coats of polyurethane	"	2.10	1.16	3.26
9540	Clean and wax floors	"	0.31	0.24	0.55
09550.20	**BAMBOO FLOORING**				
0010	Vertical, Carbonized Medium, 3' vertical grain	S.F.	2.10	5.85	7.95
0020	Natural	"	2.10	5.75	7.85
0030	3' horizontal grain	"	2.10	5.85	7.95
0040	Natural	"	2.10	5.75	7.85
0050	3' Stained	"	2.10	5.64	7.74
0060	6' spice	"	2.10	5.64	7.74
0070	3' stained, butterscotch	"	2.10	5.64	7.74
0080	6' tiger	"	2.10	5.64	7.74
0090	3' stained, Irish moss	"	2.10	5.64	7.74
0100	Vice-lock, 12 mm., laminate flooring, maple	"	2.10	3.72	5.82
0101	Oak	"	2.10	3.20	5.30
0102	Pine	"	2.10	4.26	6.36
0103	Espresso	"	2.10	4.26	6.36
0104	Standard, hard maple	"	2.10	3.46	5.56
0105	Cherry	"	2.10	3.20	5.30
0106	Oak	"	2.10	2.13	4.23
0107	Walnut	"	2.10	2.13	4.23
0108	South pacific vice-lock, 12 mm, brazilian cherry	"	2.10	4.79	6.89
0109	Maple	"	2.10	4.26	6.36
0110	Teak	"	2.10	4.53	6.63
09630.10	**UNIT MASONRY FLOORING**				
1000	Clay brick				
1020	9 x 4-1/2 x 3" thick				
1040	Glazed	S.F.	5.25	8.03	13.28
1060	Unglazed	"	5.25	7.70	12.95
1070	8 x 4 x 3/4" thick				
1080	Glazed	S.F.	5.47	7.26	12.73
1100	Unglazed	"	5.47	6.93	12.40
1140	For herringbone pattern, add to labor, 15%				
09660.10	**RESILIENT TILE FLOORING**				
1020	Solid vinyl tile, 1/8" thick, 12" x 12"				
1040	Marble patterns	S.F.	1.57	4.66	6.23
1060	Solid colors	"	1.57	6.05	7.62
1080	Travertine patterns	"	1.57	6.79	8.36
2000	Conductive resilient flooring, vinyl tile				

		UNIT	LABOR	MAT.	TOTAL
09660.10	**RESILIENT TILE FLOORING, Cont'd...**				
2040	1/8" thick, 12" x 12"	S.F.	1.80	7.38	9.18
09665.10	**RESILIENT SHEET FLOORING**				
0980	Vinyl sheet flooring				
1000	Minimum	S.F.	0.63	3.83	4.46
1002	Average	"	0.75	6.19	6.94
1004	Maximum	"	1.05	10.50	11.55
1020	Cove, to 6"	L.F.	1.26	2.28	3.54
2000	Fluid applied resilient flooring				
2020	Polyurethane, poured in place, 3/8" thick	S.F.	5.25	10.50	15.75
6200	Vinyl sheet goods, backed				
6220	0.070" thick	S.F.	0.78	3.90	4.68
6260	0.125" thick	"	0.78	6.98	7.76
6280	0.250" thick	"	0.78	8.03	8.81
09678.10	**RESILIENT BASE AND ACCESSORIES**				
1000	Wall base, vinyl				
1130	4" high	L.F.	2.10	1.28	3.38
1140	6" high	"	2.10	1.74	3.84
09682.10	**CARPET PADDING**				
1000	Carpet padding				
1005	Foam rubber, waffle type, 0.3" thick	S.Y.	3.15	6.74	9.89
1010	Jute padding				
1022	Average	S.Y.	3.15	5.95	9.10
1030	Sponge rubber cushion				
1042	Average	S.Y.	3.15	7.22	10.37
1050	Urethane cushion, 3/8" thick				
1062	Average	S.Y.	3.15	6.32	9.47
09685.10	**CARPET**				
0990	Carpet, acrylic				
1000	24 oz., light traffic	S.Y.	7.00	17.75	24.75
1020	28 oz., medium traffic	"	7.00	21.25	28.25
2010	Nylon				
2020	15 oz., light traffic	S.Y.	7.00	24.50	31.50
2040	28 oz., medium traffic	"	7.00	32.00	39.00
2110	Nylon				
2120	28 oz., medium traffic	S.Y.	7.00	30.50	37.50
2140	35 oz., heavy traffic	"	7.00	37.25	44.25
2145	Wool				
2150	30 oz., medium traffic	S.Y.	7.00	51.00	58.00
2160	36 oz., medium traffic	"	7.00	53.00	60.00
2180	42 oz., heavy traffic	"	7.00	71.00	78.00
3000	Carpet tile				
3020	Foam backed				
3022	Minimum	S.F.	1.26	4.07	5.33
3024	Average	"	1.40	4.71	6.11
3026	Maximum	"	1.57	7.47	9.04
8980	Clean and vacuum carpet				
9000	Minimum	S.Y.	0.24	0.36	0.60
9020	Average	"	0.42	0.56	0.98
9040	Maximum	"	0.63	0.77	1.40

		UNIT	LABOR	MAT.	TOTAL
09905.10	**PAINTING PREPARATION**				
1000	Dropcloths				
1050	Minimum	S.F.	0.03	0.03	0.06
1100	Average	"	0.04	0.06	0.10
1150	Maximum	"	0.05	0.07	0.12
1200	Masking				
1250	Paper and tape				
1300	Minimum	L.F.	0.52	0.02	0.54
1350	Average	"	0.65	0.03	0.68
1400	Maximum	"	0.87	0.05	0.92
1450	Doors				
1500	Minimum	EA.	6.57	0.05	6.62
1550	Average	"	8.76	0.06	8.82
1600	Maximum	"	11.75	0.07	11.82
1650	Windows				
1700	Minimum	EA.	6.57	0.05	6.62
1750	Average	"	8.76	0.06	8.82
1800	Maximum	"	11.75	0.07	11.82
2000	Sanding				
2050	Walls and flat surfaces				
2100	Minimum	S.F.	0.35		0.35
2150	Average	"	0.43		0.43
2200	Maximum	"	0.52		0.52
2250	Doors and windows				
2300	Minimum	EA.	8.76		8.76
2350	Average	"	13.25		13.25
2400	Maximum	"	17.50		17.50
2450	Trim				
2500	Minimum	L.F.	0.65		0.65
2550	Average	"	0.87		0.87
2600	Maximum	"	1.16		1.16
2650	Puttying				
2700	Minimum	S.F.	0.80	0.01	0.81
2750	Average	"	1.05	0.02	1.07
2800	Maximum	"	1.31	0.03	1.34
09910.05	**EXT. PAINTING, SITEWORK**				
3000	Concrete Block				
3020	Roller				
3040	First Coat				
3060	Minimum	S.F.	0.26	0.19	0.45
3080	Average	"	0.35	0.21	0.56
3100	Maximum	"	0.52	0.22	0.74
3120	Second Coat				
3140	Minimum	S.F.	0.21	0.19	0.40
3160	Average	"	0.29	0.21	0.50
3180	Maximum	"	0.43	0.22	0.65
3200	Spray				
3220	First Coat				
3240	Minimum	S.F.	0.14	0.15	0.29
3260	Average	"	0.17	0.17	0.34
3280	Maximum	"	0.20	0.18	0.38
3300	Second Coat				

		UNIT	LABOR	MAT.	TOTAL
09910.05	**EXT. PAINTING, SITEWORK, Cont'd...**				
3320	Minimum	S.F.	0.09	0.15	0.24
3340	Average	"	0.11	0.17	0.28
3360	Maximum	"	0.16	0.18	0.34
3500	Fences, Chain Link				
3700	Roller				
3720	First Coat				
3740	Minimum	S.F.	0.37	0.13	0.50
3760	Average	"	0.43	0.14	0.57
3780	Maximum	"	0.50	0.15	0.65
3800	Second Coat				
3820	Minimum	S.F.	0.21	0.13	0.34
3840	Average	"	0.26	0.14	0.40
3860	Maximum	"	0.32	0.15	0.47
3880	Spray				
3900	First Coat				
3920	Minimum	S.F.	0.16	0.10	0.26
3940	Average	"	0.18	0.11	0.29
3960	Maximum	"	0.21	0.13	0.34
3980	Second Coat				
4000	Minimum	S.F.	0.12	0.10	0.22
4060	Average	"	0.14	0.11	0.25
4080	Maximum	"	0.16	0.13	0.29
4200	Fences, Wood or Masonry				
4220	Brush				
4240	First Coat				
4260	Minimum	S.F.	0.55	0.19	0.74
4280	Average	"	0.65	0.21	0.86
4300	Maximum	"	0.87	0.22	1.09
4320	Second Coat				
4340	Minimum	S.F.	0.32	0.19	0.51
4360	Average	"	0.40	0.21	0.61
4380	Maximum	"	0.52	0.22	0.74
4400	Roller				
4420	First Coat				
4440	Minimum	S.F.	0.29	0.19	0.48
4460	Average	"	0.35	0.21	0.56
4480	Maximum	"	0.40	0.22	0.62
4500	Second Coat				
4520	Minimum	S.F.	0.20	0.19	0.39
4540	Average	"	0.25	0.21	0.46
4560	Maximum	"	0.32	0.22	0.54
4580	Spray				
4600	First Coat				
4620	Minimum	S.F.	0.18	0.15	0.33
4640	Average	"	0.23	0.17	0.40
4660	Maximum	"	0.32	0.18	0.50
4680	Second Coat				
4700	Minimum	S.F.	0.13	0.15	0.28
4760	Average	"	0.16	0.17	0.33
4780	Maximum	"	0.21	0.18	0.39

DIVISION # 09 FINISHES

09910.15	EXT. PAINTING, BUILDINGS	UNIT	LABOR	MAT.	TOTAL
1200	Decks, Wood, Stained				
1580	Spray				
1600	First Coat				
1620	Minimum	S.F.	0.16	0.13	0.29
1640	Average	"	0.17	0.14	0.31
1660	Maximum	"	0.20	0.15	0.35
1680	Second Coat				
1700	Minimum	S.F.	0.14	0.13	0.27
1720	Average	"	0.15	0.14	0.29
1740	Maximum	"	0.17	0.15	0.32
2520	Doors, Wood				
2540	Brush				
2560	First Coat				
2580	Minimum	S.F.	0.80	0.15	0.95
2600	Average	"	1.05	0.17	1.22
2620	Maximum	"	1.31	0.18	1.49
2640	Second Coat				
2660	Minimum	S.F.	0.65	0.15	0.80
2680	Average	"	0.75	0.17	0.92
2700	Maximum	"	0.87	0.18	1.05
3680	Siding, Wood				
3880	Spray				
3900	First Coat				
3920	Minimum	S.F.	0.17	0.13	0.30
3940	Average	"	0.18	0.14	0.32
3960	Maximum	"	0.20	0.15	0.35
3980	Second Coat				
4000	Minimum	S.F.	0.13	0.13	0.26
4020	Average	"	0.17	0.14	0.31
4040	Maximum	"	0.26	0.15	0.41
4440	Trim				
4460	Brush				
4480	First Coat				
4500	Minimum	L.F.	0.21	0.19	0.40
4520	Average	"	0.26	0.21	0.47
4540	Maximum	"	0.32	0.22	0.54
4560	Second Coat				
4580	Minimum	L.F.	0.16	0.19	0.35
4600	Average	"	0.21	0.21	0.42
4620	Maximum	"	0.32	0.22	0.54
4640	Walls				
4840	Spray				
4860	First Coat				
4880	Minimum	S.F.	0.08	0.11	0.19
4900	Average	"	0.10	0.13	0.23
4920	Maximum	"	0.13	0.14	0.27
4940	Second Coat				
4960	Minimum	S.F.	0.07	0.11	0.18
4980	Average	"	0.08	0.13	0.21
5000	Maximum	"	0.11	0.14	0.25
5020	Windows				
5040	Brush				

		UNIT	LABOR	MAT.	TOTAL
09910.15	**EXT. PAINTING, BUILDINGS, Cont'd...**				
5060	First Coat				
5080	Minimum	S.F.	0.87	0.13	1.00
5100	Average	"	1.05	0.14	1.19
5120	Maximum	"	1.31	0.15	1.46
5140	Second Coat				
5160	Minimum	S.F.	0.75	0.13	0.88
5180	Average	"	0.87	0.14	1.01
5200	Maximum	"	1.05	0.15	1.20
09910.35	**INT. PAINTING, BUILDINGS**				
1380	Cabinets and Casework				
1400	Brush				
1420	First Coat				
1440	Minimum	S.F.	0.52	0.19	0.71
1460	Average	"	0.58	0.21	0.79
1480	Maximum	"	0.65	0.22	0.87
1500	Second Coat				
1520	Minimum	S.F.	0.43	0.19	0.62
1540	Average	"	0.47	0.21	0.68
1560	Maximum	"	0.52	0.22	0.74
1580	Spray				
1600	First Coat				
1620	Minimum	S.F.	0.26	0.15	0.41
1640	Average	"	0.30	0.17	0.47
1660	Maximum	"	0.37	0.18	0.55
1680	Second Coat				
1700	Minimum	S.F.	0.21	0.15	0.36
1720	Average	"	0.22	0.17	0.39
1740	Maximum	"	0.29	0.18	0.47
2520	Doors, Wood				
2540	Brush				
2560	First Coat				
2580	Minimum	S.F.	0.75	0.19	0.94
2600	Average	"	0.95	0.21	1.16
2620	Maximum	"	1.16	0.22	1.38
2640	Second Coat				
2660	Minimum	S.F.	0.58	0.14	0.72
2680	Average	"	0.65	0.15	0.80
2700	Maximum	"	0.75	0.17	0.92
2720	Spray				
2740	First Coat				
2760	Minimum	S.F.	0.15	0.14	0.29
2780	Average	"	0.18	0.15	0.33
2800	Maximum	"	0.23	0.17	0.40
2820	Second Coat				
2840	Minimum	S.F.	0.12	0.14	0.26
2860	Average	"	0.14	0.15	0.29
2880	Maximum	"	0.16	0.17	0.33
3900	Trim				
3920	Brush				
3940	First Coat				
3960	Minimum	L.F.	0.21	0.19	0.40

		UNIT	LABOR	MAT.	TOTAL
09910.35	**INT. PAINTING, BUILDINGS, Cont'd...**				
3980	Average	L.F.	0.23	0.21	0.44
4000	Maximum	"	0.29	0.22	0.51
4020	Second Coat				
4040	Minimum	L.F.	0.15	0.19	0.34
4060	Average	"	0.20	0.21	0.41
4080	Maximum	"	0.29	0.22	0.51
4100	Walls				
4120	Roller				
4140	First Coat				
4160	Minimum	S.F.	0.18	0.15	0.33
4180	Average	"	0.19	0.17	0.36
4200	Maximum	"	0.21	0.18	0.39
4220	Second Coat				
4240	Minimum	S.F.	0.16	0.15	0.31
4260	Average	"	0.17	0.17	0.34
4280	Maximum	"	0.20	0.18	0.38
4300	Spray				
4320	First Coat				
4340	Minimum	S.F.	0.08	0.13	0.21
4360	Average	"	0.10	0.14	0.24
4380	Maximum	"	0.13	0.15	0.28
4400	Second Coat				
4420	Minimum	S.F.	0.07	0.13	0.20
4440	Average	"	0.09	0.14	0.23
4460	Maximum	"	0.11	0.15	0.26
09955.10	**WALL COVERING**				
0900	Vinyl wall covering				
1000	Medium duty	S.F.	0.75	0.95	1.70
1010	Heavy duty	"	0.87	1.96	2.83

DCD

Design Cost Data™

TABLE OF CONTENTS   PAGE

		UNIT	LABOR	MAT.	TOTAL
10110.10	**CHALKBOARDS**				
1020	Chalkboard, metal frame, 1/4" thick				
1040	48"x60"	EA.	63.00	470	533
1060	48"x96"	"	70.00	650	720
1080	48"x144"	"	79.00	860	939
1100	48"x192"	"	90.00	1,170	1,260
1110	Liquid chalkboard				
1120	48"x60"	EA.	63.00	630	693
1140	48"x96"	"	70.00	800	870
1160	48"x144"	"	79.00	1,200	1,279
1180	48"x192"	"	90.00	1,370	1,460
1200	Map rail, deluxe	L.F.	3.15	7.82	10.97
10165.10	**TOILET PARTITIONS**				
0100	Toilet partition, plastic laminate				
0120	Ceiling mounted	EA.	210	1,160	1,370
0140	Floor mounted	"	160	760	920
0150	Metal				
0165	Ceiling mounted	EA.	210	790	1,000
0180	Floor mounted	"	160	750	910
0190	Wheel chair partition, plastic laminate				
0200	Ceiling mounted	EA.	210	1,730	1,940
0210	Floor mounted	"	160	1,520	1,680
0215	Painted metal				
0220	Ceiling mounted	EA.	210	1,240	1,450
0230	Floor mounted	"	160	1,120	1,280
1980	Urinal screen, plastic laminate				
2000	Wall hung	EA.	79.00	530	609
2100	Floor mounted	"	79.00	480	559
2120	Porcelain enameled steel, floor mounted	"	79.00	620	699
2140	Painted metal, floor mounted	"	79.00	410	489
2160	Stainless steel, floor mounted	"	79.00	780	859
5000	Metal toilet partitions				
5020	Front door and side divider, floor mounted				
5040	Porcelain enameled steel	EA.	160	1,270	1,430
5060	Painted steel	"	160	750	910
5080	Stainless steel	"	160	1,860	2,020
10185.10	**SHOWER STALLS**				
1000	Shower receptors				
1010	Precast, terrazzo				
1020	32" x 32"	EA.	58.00	670	728
1040	32" x 48"	"	69.00	710	779
1050	Concrete				
1060	32" x 32"	EA.	58.00	280	338
1080	48" x 48"	"	77.00	310	387
1100	Shower door, trim and hardware				
1130	Porcelain enameled steel, flush	EA.	69.00	550	619
1140	Baked enameled steel, flush	"	69.00	330	399
1150	Aluminum, tempered glass, 48" wide, sliding	"	86.00	680	766
1161	Folding	"	86.00	650	736
5400	Shower compartment, precast concrete receptor				
5420	Single entry type				
5440	Porcelain enameled steel	EA.	690	2,330	3,020

		UNIT	LABOR	MAT.	TOTAL
10185.10	**SHOWER STALLS, Cont'd...**				
5460	Baked enameled steel	EA.	690	2,240	2,930
5480	Stainless steel	"	690	2,150	2,840
5500	Double entry type				
5520	Porcelain enameled steel	EA.	860	4,160	5,020
5540	Baked enameled steel	"	860	2,840	3,700
5560	Stainless steel	"	860	4,600	5,460
10210.10	**VENTS AND WALL LOUVERS**				
1230	Grilles and louvers				
2040	Fixed type louvers				
2060	4 through 10 sf	S.F.	11.50	33.75	45.25
2080	Over 10 sf	"	8.70	40.00	48.70
2090	Movable type louvers				
2220	4 through 10 sf	S.F.	11.50	40.00	51.50
2240	Over 10 sf	"	8.70	44.25	52.95
2260	Aluminum louvers				
4000	Residential use, fixed type, with screen				
4020	8" x 8"	EA.	34.75	20.50	55.25
4060	12" x 12"	"	34.75	22.50	57.25
4080	12" x 18"	"	34.75	27.00	61.75
4100	14" x 24"	"	34.75	38.75	73.50
4120	18" x 24"	"	34.75	43.50	78.25
4140	30" x 24"	"	38.75	59.00	97.75
10290.10	**PEST CONTROL**				
1000	Termite control				
1010	Under slab spraying				
1020	Minimum	S.F.	0.12	1.19	1.31
1040	Average	"	0.24	1.19	1.43
1120	Maximum	"	0.49	1.70	2.19
10350.10	**FLAGPOLES**				
2020	Installed in concrete base				
2030	Fiberglass				
2040	25' high	EA.	420	1,610	2,030
2080	50' high	"	1,050	4,250	5,300
2100	Aluminum				
2120	25' high	EA.	420	1,560	1,980
2140	50' high	"	1,050	3,090	4,140
2160	Bonderized steel				
2180	25' high	EA.	480	1,750	2,230
2200	50' high	"	1,260	3,490	4,750
2220	Freestanding tapered, fiberglass				
2240	30' high	EA.	450	1,910	2,360
2260	40' high	"	570	2,490	3,060
2280	50' high	"	630	6,340	6,970
2300	60' high	"	740	6,770	7,510
2400	Wall mounted, with collar, brushed aluminum finish				
2420	15' long	EA.	310	1,480	1,790
2440	18' long	"	310	1,670	1,980
2460	20' long	"	330	1,830	2,160
2480	24' long	"	370	1,960	2,330
2500	Outrigger, wall, including base				
2520	10' long	EA.	420	1,500	1,920

		UNIT	LABOR	MAT.	TOTAL
10350.10	**FLAGPOLES, Cont'd...**				
2540	20' long	EA.	530	1,990	2,520
10400.10	**IDENTIFYING DEVICES**				
1000	Directory and bulletin boards				
1020	Open face boards				
1040	Chrome plated steel frame	S.F.	31.50	37.00	68.50
1060	Aluminum framed	"	31.50	64.00	95.50
1080	Bronze framed	"	31.50	82.00	114
1100	Stainless steel framed	"	31.50	110	142
1140	Tack board, aluminum framed	"	31.50	26.00	57.50
1160	Visual aid board, aluminum framed	"	31.50	26.00	57.50
1200	Glass encased boards, hinged and keyed				
1210	Aluminum framed	S.F.	79.00	140	219
1220	Bronze framed	"	79.00	160	239
1230	Stainless steel framed	"	79.00	210	289
1240	Chrome plated steel framed	"	79.00	220	299
2020	Metal plaque				
2040	Cast bronze	S.F.	53.00	630	683
2060	Aluminum	"	53.00	360	413
2080	Metal engraved plaque				
2100	Porcelain steel	S.F.	53.00	760	813
2120	Stainless steel	"	53.00	600	653
2140	Brass	"	53.00	900	953
2160	Aluminum	"	53.00	560	613
2200	Metal built-up plaque				
2220	Bronze	S.F.	63.00	690	753
2240	Copper and bronze	"	63.00	600	663
2260	Copper and aluminum	"	63.00	670	733
2280	Metal nameplate plaques				
2300	Cast bronze	S.F.	39.25	680	719
2320	Cast aluminum	"	39.25	500	539
2330	Engraved, 1-1/2" x 6"				
2340	Bronze	EA.	39.25	290	329
2360	Aluminum	"	39.25	220	259
2440	Letters, on masonry, aluminum, satin finish				
2450	1/2" thick				
2460	2" high	EA.	25.25	26.50	51.75
2480	4" high	"	31.50	39.75	71.25
2500	6" high	"	35.00	53.00	88.00
2510	3/4" thick				
2520	8" high	EA.	39.25	79.00	118
2540	10" high	"	45.00	92.00	137
2550	1" thick				
2560	12" high	EA.	53.00	100	153
2580	14" high	"	63.00	120	183
2600	16" high	"	79.00	140	219
2620	For polished aluminum add, 15%				
2640	For clear anodized aluminum add, 15%				
2660	For colored anodic aluminum add, 30%				
2680	For profiled and color enameled letters add, 50%				
2700	Cast bronze, satin finish letters				
2710	3/8" thick				

		UNIT	LABOR	MAT.	TOTAL
10400.10	**IDENTIFYING DEVICES, Cont'd...**				
2720	2" high	EA.	25.25	32.25	57.50
2740	4" high	"	31.50	48.25	79.75
2760	1/2" thick, 6" high	"	35.00	66.00	101
2780	5/8" thick, 8" high	"	39.25	98.00	137
2785	1" thick				
2790	10" high	EA.	45.00	120	165
2800	12" high	"	53.00	150	203
2820	14" high	"	63.00	180	243
2840	16" high	"	79.00	270	349
3000	Interior door signs, adhesive, flexible				
3060	2" x 8"	EA.	12.25	24.75	37.00
3080	4" x 4"	"	12.25	26.25	38.50
3100	6" x 7"	"	12.25	33.00	45.25
3120	6" x 9"	"	12.25	42.00	54.25
3140	10" x 9"	"	12.25	55.00	67.25
3160	10" x 12"	"	12.25	72.00	84.25
3200	Hard plastic type, no frame				
3220	3" x 8"	EA.	12.25	55.00	67.25
3240	4" x 4"	"	12.25	55.00	67.25
3260	4" x 12"	"	12.25	59.00	71.25
3280	Hard plastic type, with frame				
3300	3" x 8"	EA.	12.25	170	182
3320	4" x 4"	"	12.25	130	142
3340	4" x 12"	"	12.25	200	212
10450.10	**CONTROL**				
1020	Access control, 7' high, indoor or outdoor impenetrability				
1040	Remote or card control, type B	EA.	860	1,810	2,670
1060	Free passage, type B	"	860	1,470	2,330
1080	Remote or card control, type AA	"	860	2,890	3,750
1100	Free passage, type AA	"	860	2,620	3,480
10500.10	**LOCKERS**				
0080	Locker bench, floor mounted, laminated maple				
0100	4'	EA.	53.00	360	413
0120	6'	"	53.00	510	563
0130	Wardrobe locker, 12" x 60" x 15", baked on enamel				
0140	1-tier	EA.	31.50	390	422
0160	2-tier	"	31.50	410	442
0180	3-tier	"	33.25	460	493
0200	4-tier	"	33.25	500	533
0240	12" x 72" x 15", baked on enamel				
0260	1-tier	EA.	31.50	330	362
0280	2-tier	"	31.50	400	432
0300	4-tier	"	33.25	500	533
0320	5-tier	"	33.25	500	533
1200	15" x 60" x 15", baked on enamel				
1220	1-tier	EA.	31.50	440	472
1240	4-tier	"	33.25	470	503
10520.10	**FIRE PROTECTION**				
1000	Portable fire extinguishers				
1020	Water pump tank type				
1030	2.5 gal.				

		UNIT	LABOR	MAT.	TOTAL
10520.10	**FIRE PROTECTION, Cont'd...**				
1040	Red enameled galvanized	EA.	33.00	140	173
1060	Red enameled copper	"	33.00	220	253
1080	Polished copper	"	33.00	280	313
1200	Carbon dioxide type, red enamel steel				
1210	Squeeze grip with hose and horn				
1220	2.5 lb	EA.	33.00	220	253
1240	5 lb	"	38.00	310	348
1260	10 lb	"	49.25	330	379
1280	15 lb	"	62.00	360	422
1300	20 lb	"	62.00	450	512
1310	Wheeled type				
1320	125 lb	EA.	99.00	3,980	4,079
1340	250 lb	"	99.00	5,030	5,129
1360	500 lb	"	99.00	6,490	6,589
1400	Dry chemical, pressurized type				
1405	Red enameled steel				
1410	2.5 lb	EA.	33.00	69.00	102
1430	5 lb	"	38.00	94.00	132
1440	10 lb	"	49.25	200	249
1450	20 lb	"	62.00	250	312
1460	30 lb	"	62.00	310	372
1480	Chrome plated steel, 2.5 lb	"	33.00	300	333
10550.10	**POSTAL SPECIALTIES**				
1500	Mail chutes				
1520	Single mail chute				
1530	Finished aluminum	L.F.	160	850	1,010
1540	Bronze	"	160	1,180	1,340
1560	Single mail chute receiving box				
1580	Finished aluminum	EA.	310	1,260	1,570
1600	Bronze	"	310	1,510	1,820
1620	Twin mail chute, double parallel				
1630	Finished aluminum	FLR	310	1,850	2,160
1640	Bronze	"	310	2,420	2,730
10670.10	**SHELVING**				
0980	Shelving, enamel, closed side and back, 12" x 36"				
1000	5 shelves	EA.	110	300	410
1020	8 shelves	"	140	320	460
1030	Open				
1040	5 shelves	EA.	110	150	260
1060	8 shelves	"	140	170	310
2000	Metal storage shelving, baked enamel				
2030	7 shelf unit, 72" or 84" high				
2050	12" shelf	L.F.	66.00	47.75	114
2080	24" shelf	"	79.00	98.00	177
2100	36" shelf	"	90.00	110	200
2200	4 shelf unit, 40" high				
2240	12" shelf	L.F.	57.00	63.00	120
2270	24" shelf	"	70.00	130	200
2300	3 shelf unit, 32" high				
2340	12" shelf	L.F.	33.25	50.00	83.25
2370	24" shelf	"	39.25	63.00	102

		UNIT	LABOR	MAT.	TOTAL
10670.10	**SHELVING, Cont'd...**				
2400	Single shelf unit, attached to masonry				
2420	12" shelf	L.F.	11.50	19.50	31.00
2450	24" shelf	"	13.75	27.75	41.50
2460	For stainless steel, add to material, 120%				
2470	For attachment to gypsum board, add to labor, 50%				
10800.10	**BATH ACCESSORIES**				
1040	Ash receiver, wall mounted, aluminum	EA.	31.50	160	192
1050	Grab bar, 1-1/2" dia., stainless steel, wall mounted				
1060	24" long	EA.	31.50	53.00	84.50
1080	36" long	"	33.25	59.00	92.25
1100	48" long	"	37.00	72.00	109
1130	1" dia., stainless steel				
1140	12" long	EA.	27.50	32.75	60.25
1180	24" long	"	31.50	44.75	76.25
1220	36" long	"	35.00	59.00	94.00
1240	48" long	"	37.00	66.00	103
1300	Hand dryer, surface mounted, 110 volt	"	79.00	750	829
1320	Medicine cabinet, 16 x 22, baked enamel, lighted	"	25.25	150	175
1340	With mirror, lighted	"	42.00	220	262
1420	Mirror, 1/4" plate glass, up to 10 sf	S.F.	6.30	11.50	17.80
1430	Mirror, stainless steel frame				
1440	18"x24"	EA.	21.00	88.00	109
1500	24"x30"	"	31.50	110	142
1520	24"x48"	"	53.00	160	213
1560	30"x30"	"	63.00	350	413
1600	48"x72"	"	110	690	800
1640	With shelf, 18"x24"	"	25.25	270	295
1820	Sanitary napkin dispenser, stainless steel	"	42.00	650	692
1830	Shower rod, 1" diameter				
1840	Chrome finish over brass	EA.	31.50	240	272
1860	Stainless steel	"	31.50	160	192
1900	Soap dish, stainless steel, wall mounted	"	42.00	150	192
1910	Toilet tissue dispenser, stainless, wall mounted				
1920	Single roll	EA.	15.75	74.00	89.75
1940	Double roll	"	18.00	140	158
1945	Towel dispenser, stainless steel				
1950	Flush mounted	EA.	35.00	280	315
1960	Surface mounted	"	31.50	400	432
1970	Combination towel and waste receptacle	"	42.00	610	652
2000	Towel bar, stainless steel				
2020	18" long	EA.	25.25	91.00	116
2040	24" long	"	28.75	120	149
2060	30" long	"	31.50	130	162
2070	36" long	"	35.00	140	175
2100	Waste receptacle, stainless steel, wall mounted	"	53.00	460	513

DCD Design Cost Data™

TABLE OF CONTENTS   PAGE

DIVISION # 11 EQUIPMENT

		UNIT	LABOR	MAT.	TOTAL
11010.10	**MAINTENANCE EQUIPMENT**				
1000	Vacuum cleaning system				
1010	3 valves				
1020	1.5 hp	EA.	700	980	1,680
1030	2.5 hp	"	900	1,170	2,070
1040	5 valves	"	1,260	1,830	3,090
1060	7 valves	"	1,580	2,440	4,020
11060.10	**THEATER EQUIPMENT**				
1000	Roll out stage, steel frame, wood floor				
1020	Manual	S.F.	3.93	56.00	59.93
1040	Electric	"	6.30	54.00	60.30
1100	Portable stages				
1120	8" high	S.F.	3.15	22.25	25.40
1140	18" high	"	3.50	25.75	29.25
1160	36" high	"	3.70	30.00	33.70
1180	48" high	"	3.93	33.75	37.68
1300	Band risers				
1320	Minimum	S.F.	3.15	58.00	61.15
1340	Maximum	"	3.15	120	123
1400	Chairs for risers				
1420	Minimum	EA.	2.24	750	752
1440	Maximum	"	2.24	1,200	1,202
2000	Theatre controlls				
2010	Fade console, 48 channel	EA.	320	3,000	3,320
2020	Light control modules, 125 channels	"	640	8,530	9,170
2030	Dimmer module, stage-pin output	"	320	2,620	2,940
2040	6-Module pack, w/24 U-ground connectors	"	320	7,500	7,820
11090.10	**CHECKROOM EQUIPMENT**				
1000	Motorized checkroom equipment				
1020	No shelf system, 6'4" height				
1040	7'6" length	EA.	630	5,310	5,940
1060	14'6" length	"	630	5,370	6,000
1080	28' length	"	630	6,470	7,100
1100	One shelf, 6'8" height				
1120	7'6" length	EA.	630	6,510	7,140
1140	14'6" length	"	630	6,680	7,310
1160	28' length	"	630	8,040	8,670
11110.10	**LAUNDRY EQUIPMENT**				
1000	High capacity, heavy duty				
1020	Washer extractors				
1030	135 lb				
1040	Standard	EA.	530	36,460	36,990
1060	Pass through	"	530	41,890	42,420
1070	200 lb				
1080	Standard	EA.	530	44,770	45,300
1100	Pass through	"	530	54,370	54,900
1120	110 lb dryer	"	530	13,710	14,240
11161.10	**LOADING DOCK EQUIPMENT**				
0080	Dock leveler, 10 ton capacity				
0100	6' x 8'	EA.	630	5,650	6,280
0120	7' x 8'	"	630	6,480	7,110

11 - 2

© 2018 By Design & Construction Resources

		UNIT	LABOR	MAT.	TOTAL
11170.10	**WASTE HANDLING**				
1500	Industrial compactor				
1520	1 cy	EA.	710	13,970	14,680
1540	3 cy	"	920	21,790	22,710
1560	5 cy	"	1,280	41,450	42,730
11170.20	**AERATION EQUIPMENT**				
0010	Surface spray/Vertical pump				
0020	1 hp. Pump, 500 gpm.	EA.	350	5,200	5,550
0030	5 hp. Pump, 2000 gpm.	"	1,380	7,790	9,170
0040	Polycarbon, treatment container, 1,000 gallon capacity	"	2,720	4,140	6,860
0050	1,500 gallon	"	2,720	4,750	7,470
0060	2,000 gallon	"	3,400	6,620	10,020
0070	3,000 gallon	"	3,630	10,530	14,160
11400.10	**FOOD SERVICE EQUIPMENT**				
1000	Unit kitchens				
1020	30" compact kitchen				
1040	Refrigerator, with range, sink	EA.	320	1,510	1,830
1060	Sink only	"	210	1,920	2,130
1080	Range only	"	160	1,560	1,720
1100	Cabinet for upper wall section	"	92.00	390	482
1120	Stainless shield, for rear wall	"	25.75	160	186
1140	Side wall	"	25.75	120	146
1200	42" compact kitchen				
1220	Refrigerator with range, sink	EA.	360	1,840	2,200
1240	Sink only	"	320	1,000	1,320
1260	Cabinet for upper wall section	"	110	780	890
1280	Stainless shield, for rear wall	"	26.75	630	657
1290	Side wall	"	26.75	170	197
1600	Bake oven				
1610	Single deck				
1620	Minimum	EA.	80.00	3,560	3,640
1640	Maximum	"	160	6,810	6,970
1650	Double deck				
1660	Minimum	EA.	110	6,340	6,450
1680	Maximum	"	160	19,820	19,980
1685	Triple deck				
1690	Minimum	EA.	110	22,490	22,600
1695	Maximum	"	210	40,120	40,330
1700	Convection type oven, electric, 40" x 45" x 57"				
1720	Minimum	EA.	80.00	3,510	3,590
1740	Maximum	"	160	6,180	6,340
1800	Broiler, without oven, 69" x 26" x 39"				
1820	Minimum	EA.	80.00	5,590	5,670
1840	Maximum	"	110	8,790	8,900
1900	Coffee urns, 10 gallons				
1920	Minimum	EA.	210	4,260	4,470
1940	Maximum	"	320	4,810	5,130
2000	Fryer, with submerger				
2010	Single				
2020	Minimum	EA.	130	1,740	1,870
2040	Maximum	"	210	4,790	5,000
2050	Double				

		UNIT	LABOR	MAT.	TOTAL
11400.10	**FOOD SERVICE EQUIPMENT, Cont'd...**				
2060	Minimum	EA.	160	2,890	3,050
2080	Maximum	"	210	15,500	15,710
2100	Griddle, counter				
2110	3' long				
2120	Minimum	EA.	110	2,390	2,500
2140	Maximum	"	130	4,920	5,050
2150	5' long				
2160	Minimum	EA.	160	5,180	5,340
2180	Maximum	"	210	11,930	12,140
2200	Kettles, steam, jacketed				
2210	20 gallons				
2220	Minimum	EA.	160	11,390	11,550
2240	Maximum	"	320	12,450	12,770
2300	Range				
2310	Heavy duty, single oven, open top				
2320	Minimum	EA.	80.00	7,340	7,420
2340	Maximum	"	210	15,460	15,670
2350	Fry top				
2360	Minimum	EA.	80.00	7,510	7,590
2380	Maximum	"	210	10,510	10,720
2400	Steamers, electric				
2410	27 kw				
2420	Minimum	EA.	160	13,330	13,490
2440	Maximum	"	210	24,370	24,580
2450	18 kw				
2460	Minimum	EA.	160	7,340	7,500
2480	Maximum	"	210	17,210	17,420
2500	Dishwasher, rack type				
2520	Single tank, 190 racks/hr	EA.	320	19,540	19,860
2530	Double tank				
2540	234 racks/hr	EA.	360	40,350	40,710
2560	265 racks/hr	"	430	48,240	48,670
2580	Dishwasher, automatic 100 meals/hr	"	210	15,690	15,900
2600	Disposals				
2620	100 gal/hr	EA.	210	1,310	1,520
2640	120 gal/hr	"	220	1,520	1,740
2660	250 gal/hr	"	230	1,790	2,020
2680	Exhaust hood for dishwasher, gutter 4 sides				
2681	4'x4'x2'	EA.	240	3,170	3,410
2690	4'x7'x2'	"	260	4,300	4,560
2900	Ice cube maker				
2910	50 lb per day				
2920	Minimum	EA.	640	2,450	3,090
2940	Maximum	"	640	3,620	4,260
2950	500 lb per day				
2960	Minimum	EA.	1,070	5,820	6,890
2970	Maximum	"	1,070	6,880	7,950
3100	Refrigerated cases				
3120	Dairy products				
3140	Multi deck type	L.F.	42.75	1,380	1,423
3160	For rear sliding doors, add	"			260
3180	Delicatessen case, service deli				

		UNIT	LABOR	MAT.	TOTAL
11400.10	**FOOD SERVICE EQUIPMENT, Cont'd...**				
3190	Single deck	L.F.	320	990	1,310
3200	Multi deck	"	400	1,130	1,530
3220	Meat case				
3230	Single deck	L.F.	380	850	1,230
3240	Multi deck	"	400	990	1,390
3260	Produce case				
3270	Single deck	L.F.	380	980	1,360
3280	Multi deck	"	400	1,050	1,450
3300	Bottle coolers				
3310	6' long				
3320	Minimum	EA.	1,280	2,800	4,080
3340	Maximum	"	1,280	4,140	5,420
3345	10' long				
3350	Minimum	EA.	2,140	3,620	5,760
3360	Maximum	"	2,140	7,120	9,260
3420	Frozen food cases				
3440	Chest type	L.F.	380	760	1,140
3460	Reach-in, glass door	"	400	1,040	1,440
3470	Island case, single	"	380	930	1,310
3480	Multi deck	"	400	1,480	1,880
3500	Ice storage bins				
3520	500 lb capacity	EA.	920	1,560	2,480
3530	1000 lb capacity	"	1,830	2,330	4,160
11450.10	**RESIDENTIAL EQUIPMENT**				
0300	Compactor, 4 to 1 compaction	EA.	160	1,560	1,720
1310	Dishwasher, built-in				
1320	2 cycles	EA.	320	770	1,090
1330	4 or more cycles	"	320	2,070	2,390
1340	Disposal				
1350	Garbage disposer	EA.	210	210	420
1360	Heaters, electric, built-in				
1362	Ceiling type	EA.	210	440	650
1364	Wall type				
1370	Minimum	EA.	160	220	380
1374	Maximum	"	210	760	970
1400	Hood for range, 2-speed, vented				
1420	30" wide	EA.	210	610	820
1440	42" wide	"	210	1,120	1,330
1460	Ice maker, automatic				
1480	30 lb per day	EA.	92.00	2,040	2,132
1500	50 lb per day	"	320	2,590	2,910
2000	Ranges electric				
2040	Built-in, 30", 1 oven	EA.	210	2,240	2,450
2050	2 oven	"	210	2,590	2,800
2060	Counter top, 4 burner, standard	"	160	1,290	1,450
2070	With grill	"	160	3,230	3,390
2198	Free standing, 21", 1 oven	"	210	1,160	1,370
2200	30", 1 oven	"	130	2,260	2,390
2220	2 oven	"	130	3,690	3,820
3600	Water softener				
3620	30 grains per gallon	EA.	210	1,260	1,470

		UNIT	LABOR	MAT.	TOTAL
11450.10	**RESIDENTIAL EQUIPMENT, Cont'd...**				
3640	70 grains per gallon	EA.	320	1,590	1,910
11600.10	**LABORATORY EQUIPMENT**				
1000	Cabinets, base				
1020	Minimum	L.F.	53.00	480	533
1040	Maximum	"	53.00	870	923
1080	Full storage, 7' high				
1100	Minimum	L.F.	53.00	460	513
1140	Maximum	"	53.00	880	933
1150	Wall				
1160	Minimum	L.F.	63.00	170	233
1200	Maximum	"	63.00	300	363
1220	Counter tops				
1240	Minimum	S.F.	7.87	71.00	78.87
1260	Average	"	9.00	84.00	93.00
1280	Maximum	"	10.50	97.00	108
1300	Tables				
1320	Open underneath	S.F.	31.50	160	192
1330	Doors underneath	"	39.25	520	559
2000	Medical laboratory equipment				
2010	Analyzer				
2020	Chloride	EA.	32.00	5,750	5,782
2060	Blood	"	53.00	31,610	31,663
2070	Bath, water, utility, countertop unit	"	64.00	1,190	1,254
2080	Hot plate, lab, countertop	"	58.00	460	518
2100	Stirrer	"	58.00	550	608
2120	Incubator, anaerobic, 23x23x36"	"	320	9,750	10,070
2140	Dry heat bath	"	110	1,050	1,160
2160	Incinerator, for sterilizing	"	6.41	720	726
2170	Meter, serum protein	"	8.02	1,120	1,128
2180	Ph analog, general purpose	"	9.16	1,210	1,219
2190	Refrigerator, blood bank	"	110	9,220	9,330
2200	5.4 cf, undercounter type	"	110	6,260	6,370
2210	Refrigerator/freezer, 4.4 cf, undercounter type	"	110	1,190	1,300
2220	Sealer, impulse, free standing, 20x12x4"	"	21.50	660	682
2240	Timer, electric, 1-60 minutes, bench or wall mounted	"	35.75	250	286
2260	Glassware washer - dryer, undercounter	"	800	10,910	11,710
2300	Balance, torsion suspension, tabletop, 4.5 lb capacity	"	35.75	1,430	1,466
2340	Binocular microscope, with in base illuminator	"	24.75	4,380	4,405
2400	Centrifuge, table model, 19x16x13"	"	25.75	1,680	1,706
2420	Clinical model, with four place head	"	14.25	1,810	1,824
11700.10	**MEDICAL EQUIPMENT**				
1000	Hospital equipment, lights				
1020	Examination, portable	EA.	53.00	2,000	2,053
1200	Meters				
1220	Air flow meter	EA.	35.75	110	146
1240	Oxygen flow meters	"	26.75	130	157
1900	Physical therapy				
1930	Chair, hydrotherapy	EA.	10.50	780	791
1940	Diathermy, shortwave, portable, on casters	"	25.25	3,340	3,365
1950	Exercise bicycle, floor standing, 35" x 15"	"	21.00	3,240	3,261
1960	Hydrocollator, 4 pack, portable, 129 x 90 x 160"	"	9.00	580	589

		UNIT	LABOR	MAT.	TOTAL
11700.10	**MEDICAL EQUIPMENT, Cont'd...**				
1970	Lamp, infrared, mobile with variable heat control	EA.	48.50	790	839
1980	Ultra violet, base mounted	"	48.50	720	769
2070	Stimulator, galvanic-faradic, hand held	"	4.20	410	414
2080	Ultrasound, stimulator, portable, 13x13x8"	"	5.25	3,470	3,475
2120	Whirlpool, 85 gallon	"	310	6,660	6,970
2141	65 gallon capacity	"	310	6,010	6,320
2200	Radiology				
2280	Radiographic table, motor driven tilting table	EA.	6,300	58,760	65,060
2290	Fluoroscope image/tv system	"	12,600	97,490	110,090
2300	Processor for washing and drying radiographs				
2303	Water filter unit, 30" x 48-1/2" x 37-1/2"	EA.	1,070	130	1,200
2400	Steam sterilizers				
2410	For heat and moisture stable materials	EA.	64.00	5,670	5,734
2440	For fast drying after sterilization	"	80.00	7,350	7,430
2450	Compact unit	"	80.00	2,380	2,460
2460	Semi-automatic	"	320	2,800	3,120
2465	Floor loading				
2470	Single door	EA.	530	83,200	83,730
2480	Double door	"	640	91,150	91,790
2490	Utensil washer, sanitizer	"	490	18,300	18,790
2500	Automatic washer/sterilizer	"	1,280	20,030	21,310
2510	16 x 16 x 26", including accessories	"	2,140	23,020	25,160
2520	Steam generator, elec., 10 kw to 180 kw	"	1,280	38,060	39,340
2540	Surgical scrub				
2560	Minimum	EA.	210	2,000	2,210
2580	Maximum	"	210	11,630	11,840
2600	Gas sterilizers				
2620	Automatic, free standing, 21x19x29"	EA.	640	7,140	7,780
2720	Surgical lights, ceiling mounted				
2740	Minimum	EA.	1,070	9,880	10,950
2760	Maximum	"	1,280	20,170	21,450
2800	Water stills				
2900	4 liters/hr	EA.	210	4,470	4,680
2920	8 liters/hr	"	210	7,140	7,350
2940	19 liters/hr	"	530	14,690	15,220
3060	X-ray equipment				
3070	Mobile unit				
3080	Minimum	EA.	320	13,080	13,400
3100	Maximum	"	640	25,330	25,970
3200	Autopsy table				
3220	Minimum	EA.	640	18,250	18,890
3240	Maximum	"	640	25,790	26,430
3300	Incubators				
3320	15 cf	EA.	320	9,110	9,430
3330	29 cf	"	530	12,350	12,880
3340	Infant transport, portable	"	340	6,860	7,200
3460	Headwall				
3465	Aluminum, with back frame and console	EA.	320	5,140	5,460
6000	Hospital ground detection system				
6010	Power ground module	EA.	180	1,610	1,790
6020	Ground slave module	"	140	680	820
6030	Master ground module	"	120	600	720

		UNIT	LABOR	MAT.	TOTAL
11700.10	**MEDICAL EQUIPMENT, Cont'd...**				
6040	Remote indicator	EA.	130	630	760
6050	X-ray indicator	"	140	1,800	1,940
6060	Micro ammeter	"	160	2,140	2,300
6070	Supervisory module	"	140	1,800	1,940
6080	Ground cords	"	23.75	170	194
6100	Hospital isolation monitors, 5 ma				
6110	120v	EA.	280	3,410	3,690
6120	208v	"	280	3,410	3,690
6130	240v	"	280	3,690	3,970
6210	Digital clock-timers separate display	"	130	1,500	1,630
6220	One display	"	130	960	1,090
6230	Remote control	"	100	470	570
6240	Battery pack	"	100	110	210
6310	Surgical chronometer clock and 3 timers	"	200	2,800	3,000
6320	Auxiliary control	"	93.00	770	863

TABLE OF CONTENTS PAGE

		UNIT	LABOR	MAT.	TOTAL
12302.10	**CASEWORK**				
0080	Kitchen base cabinet, standard, 24" deep, 35" high				
0100	12"wide	EA.	63.00	210	273
0120	18" wide	"	63.00	250	313
0140	24" wide	"	70.00	320	390
0160	27" wide	"	70.00	360	430
0180	36" wide	"	79.00	430	509
0200	48" wide	"	79.00	520	599
0210	Drawer base, 24" deep, 35" high				
0220	15"wide	EA.	63.00	270	333
0230	18" wide	"	63.00	290	353
0240	24" wide	"	70.00	460	530
0250	27" wide	"	70.00	530	600
0260	30" wide	"	70.00	620	690
0270	Sink-ready, base cabinet				
0280	30" wide	EA.	70.00	290	360
0290	36" wide	"	70.00	300	370
0300	42" wide	"	70.00	330	400
0310	60" wide	"	79.00	390	469
0500	Corner cabinet, 36" wide	"	79.00	540	619
4000	Wall cabinet, 12" deep, 12" high				
4020	30" wide	EA.	63.00	270	333
4060	36" wide	"	63.00	290	353
4110	24" high				
4120	30" wide	EA.	70.00	360	430
4140	36" wide	"	70.00	370	440
4150	30" high				
4160	12" wide	EA.	79.00	200	279
4200	24" wide	"	79.00	250	329
4320	30" wide	"	90.00	340	430
4340	36" wide	"	90.00	340	430
4350	Corner cabinet, 30" high				
4360	24" wide	EA.	110	380	490
4390	36" wide	"	110	500	610
5020	Wardrobe	"	160	1,000	1,160
6980	Vanity with top, laminated plastic				
7000	24" wide	EA.	160	830	990
7040	36" wide	"	210	1,070	1,280
7060	48" wide	"	250	1,190	1,440
12390.10	**COUNTER TOPS**				
1020	Stainless steel, counter top, with backsplash	S.F.	15.75	250	266
2000	Acid-proof, kemrock surface	"	10.50	99.00	110
12500.10	**WINDOW TREATMENT**				
1000	Drapery tracks, wall or ceiling mounted				
1040	Basic traverse rod				
1080	50 to 90"	EA.	31.50	55.00	86.50
1100	84 to 156"	"	35.00	73.00	108
1120	136 to 250"	"	35.00	110	145
1140	165 to 312"	"	39.25	160	199
1160	Traverse rod with stationary curtain rod				
1180	30 to 50"	EA.	31.50	83.00	115
1200	50 to 90"	"	31.50	95.00	127

		UNIT	LABOR	MAT.	TOTAL
12500.10	**WINDOW TREATMENT, Cont'd...**				
1220	84 to 156"	EA.	35.00	130	165
1240	136 to 250"	"	39.25	160	199
1260	Double traverse rod				
1280	30 to 50"	EA.	31.50	97.00	129
1300	50 to 84"	"	31.50	120	152
1320	84 to 156"	"	35.00	130	165
1340	136 to 250"	"	39.25	170	209
12510.10	**BLINDS**				
0990	Venetian blinds				
1000	2" slats	S.F.	1.57	38.75	40.32
1020	1" slats	"	1.57	41.50	43.07

TABLE OF CONTENTS   PAGE

		UNIT	LABOR	MAT.	TOTAL
13056.10	**VAULTS**				
1000	Floor safes				
1010	1.0 cf	EA.	53.00	920	973
1020	1.3 cf	"	79.00	1,010	1,089
13121.10	**PRE-ENGINEERED BUILDINGS**				
1080	Pre-engineered metal building, 40'x100'				
1100	14' eave height	S.F.	5.42	8.66	14.08
1120	16' eave height	"	6.26	9.82	16.08
1140	20' eave height	"	8.14	11.00	19.14
1150	60'x100'				
1160	14' eave height	S.F.	5.42	11.00	16.42
1180	16' eave height	"	6.26	12.00	18.26
1190	20' eave height	"	8.14	13.50	21.64
1195	80'x100'				
1200	14' eave height	S.F.	5.42	8.39	13.81
1210	16' eave height	"	6.26	8.66	14.92
1220	20' eave height	"	8.14	9.68	17.82
1280	100'x100'				
1300	14' eave height	S.F.	5.42	8.18	13.60
1320	16' eave height	"	6.26	8.52	14.78
1340	20' eave height	"	8.14	9.39	17.53
1350	100'x150'				
1360	14' eave height	S.F.	5.42	7.29	12.71
1380	16' eave height	"	6.26	7.57	13.83
1400	20' eave height	"	8.14	8.11	16.25
1410	120'x150'				
1420	14' eave height	S.F.	5.42	7.71	13.13
1440	16' eave height	"	6.26	7.84	14.10
1460	20' eave height	"	8.14	8.18	16.32
1480	140'x150'				
1500	14' eave height	S.F.	5.42	7.29	12.71
1520	16' eave height	"	6.26	7.47	13.73
1540	20' eave height	"	8.14	8.11	16.25
1600	160'x200'				
1620	14' eave height	S.F.	5.42	5.62	11.04
1640	16' eave height	"	6.26	5.79	12.05
1680	20' eave height	"	8.14	6.13	14.27
1690	200'x200'				
1700	14' eave height	S.F.	5.42	4.84	10.26
1720	16' eave height	"	6.26	5.32	11.58
1740	20' eave height	"	8.14	5.66	13.80
5020	Hollow metal door and frame, 6' x 7'	EA.			1,290
5030	Sectional steel overhead door, manually operated				
5040	8' x 8'	EA.			2,120
5080	12' x 12'	"			2,820
5100	Roll-up steel door, manually operated				
5120	10' x 10'	EA.			1,650
5140	12' x 12'	"			2,950
5160	For gravity ridge ventilator with birdscreen	"			710
5161	9" throat x 10'	"			770
5181	12" throat x 10'	"			940
5200	For 20" rotary vent with damper	"			350

		UNIT	LABOR	MAT.	TOTAL
13121.10	**PRE-ENGINEERED BUILDINGS, Cont'd...**				
5220	For 4' x 3' fixed louver	EA.			250
5240	For 4' x 3' aluminum sliding window	"			220
5260	For 3' x 9' fiberglass panels	"			170
8020	Liner panel, 26 ga, painted steel	S.F.	1.74	3.10	4.84
8040	Wall panel insulated, 26 ga. steel, foam core	"	1.74	9.76	11.50
8060	Roof panel, 26 ga. painted steel	"	0.99	2.93	3.92
8080	Plastic (sky light)	"	0.99	6.61	7.60
9000	Insulation, 3-1/2" thick blanket, R11	"	0.46	1.98	2.44
13152.10	**SWIMMING POOL EQUIPMENT**				
1100	Diving boards				
1110	14' long				
1120	Aluminum	EA.	270	4,830	5,100
1140	Fiberglass	"	270	3,650	3,920
1500	Ladders, heavy duty				
1510	2 steps				
1520	Minimum	EA.	99.00	1,140	1,239
1540	Maximum	"	99.00	1,790	1,889
1550	4 steps				
1560	Minimum	EA.	120	1,220	1,340
1580	Maximum	"	120	1,960	2,080
1600	Lifeguard chair				
1620	Minimum	EA.	490	3,240	3,730
1640	Maximum	"	490	5,030	5,520
1700	Lights, underwater				
1705	12 volt, with transformer, 100 watt				
1710	Incandescent	EA.	120	240	360
1715	Halogen	"	120	210	330
1720	LED	"	120	650	770
1730	110 volt				
1740	Minimum	EA.	120	1,100	1,220
1760	Maximum	"	120	2,650	2,770
1780	Ground fault interrupter for 110 volt, each light	"	41.25	240	281
2000	Pool cover				
2020	Reinforced polyethylene	S.F.	3.79	2.34	6.13
2030	Vinyl water tube				
2040	Minimum	S.F.	3.79	1.43	5.22
2060	Maximum	"	3.79	2.14	5.93
2100	Slides with water tube				
2120	Minimum	EA.	410	1,140	1,550
2140	Maximum	"	410	24,470	24,880
13152.20	**SAUNAS**				
0010	Prefabricated, cedar siding, insulated panels, prehung door,				
0020	4'x8"x4'-8"x6'-6"	EA.			6,310
0030	5'-8"x6'-8"x6'-6"	"			7,640
0040	6'-8"x6'-8"x6'-6"	"			8,890
0050	7'-8"x7'-8"x6'-6"	"			10,740
0060	7'-8"x9'-8"x6'-6"	"			14,150
13200.10	**STORAGE TANKS**				
0080	Oil storage tank, underground, single wall, no excv.				
0090	Steel				
1000	500 gals	EA.	340	3,920	4,260

		UNIT	LABOR	MAT.	TOTAL
13200.10	**STORAGE TANKS, Cont'd...**				
1020	1,000 gals	EA.	450	5,310	5,760
1980	Fiberglass, double wall				
2000	550 gals	EA.	450	11,040	11,490
2020	1,000 gals	"	450	14,200	14,650
2520	Above ground				
2530	Steel, single wall				
2540	275 gals	EA.	270	2,220	2,490
2560	500 gals	"	450	5,550	6,000
2570	1,000 gals	"	540	7,580	8,120
3020	Fill cap	"	69.00	140	209
3040	Vent cap	"	69.00	140	209
3100	Level indicator	"	69.00	210	279

TABLE OF CONTENTS   PAGE

		UNIT	LABOR	MAT.	TOTAL
14210.10	**ELEVATORS**				
0120	Passenger elevators, electric, geared				
0502	Based on a shaft of 6 stops and 6 openings				
0510	50 fpm, 2000 lb	EA.	2,720	140,710	143,430
0520	100 fpm, 2000 lb	"	3,030	145,920	148,950
0525	150 fpm				
0530	2000 lb	EA.	3,400	160,990	164,390
0550	3000 lb	"	3,890	202,780	206,670
0560	4000 lb	"	4,540	211,000	215,540
1002	Based on a shaft of 8 stops and 8 openings				
1010	300 fpm				
1020	3000 lb	EA.	5,450	261,050	266,500
1040	3500 lb	"	5,450	265,170	270,620
1060	4000 lb	"	6,050	278,230	284,280
1080	5000 lb	"	6,490	309,140	315,630
1502	Hydraulic, based on a shaft of 3 stops, 3 openings				
1508	50 fpm				
1510	2000 lb	EA.	2,270	97,830	100,100
1520	2500 lb	"	2,270	104,610	106,880
1530	3000 lb	"	2,370	110,400	112,770
1600	For each additional; 50 fpm add per stop, $3500				
1620	500 lb, add per stop, $3500				
1630	Opening, add, $4200				
1640	Stop, add per stop, $5300				
1660	Bonderized steel door, add per opening, $400				
1670	Colored aluminum door, add per opening, $1500				
1680	Stainless steel door, add per opening, $650				
1690	Cast bronze door, add per opening, $1200				
1730	Custom cab interior, add per cab, $5000				
2000	Small elevators, 4 to 6 passenger capacity				
2005	Electric, push				
2010	2 stops	EA.	2,270	33,340	35,610
2020	3 stops	"	2,480	41,680	44,160
2030	4 stops	"	2,720	47,450	50,170
14300.10	**ESCALATORS**				
1000	Escalators				
1020	32" wide, floor to floor				
1040	12' high	EA.	4,540	174,060	178,600
1050	15' high	"	5,450	190,170	195,620
1060	18' high	"	6,810	204,820	211,630
1070	22' high	"	9,080	202,680	211,760
1080	25' high	"	10,900	230,320	241,220
1085	48" wide				
1090	12' high	EA.	4,700	193,920	198,620
1100	15' high	"	5,670	211,640	217,310
1120	18' high	"	7,170	227,350	234,520
1130	22' high	"	9,730	254,500	264,230
1140	25' high	"	10,900	271,750	282,650
14410.10	**PERSONNEL LIFTS**				
1000	Electrically operated, 1 or 2 person lift				
1001	With attached foot platforms				
1020	3 stops	EA.			11,970

		UNIT	LABOR	MAT.	TOTAL
14410.10	**PERSONNEL LIFTS, Cont'd...**				
1040	5 stops	EA.			18,670
1060	7 stops	"			21,760
2000	For each additional stop, add $1250				
3020	Residential stair climber, per story	EA.	530	5,550	6,080
14450.10	**VEHICLE LIFTS**				
1020	Automotive hoist, one post, semi-hydraulic, 8,000 lb	EA.	2,720	4,000	6,720
1040	Full hydraulic, 8,000 lb	"	2,720	4,120	6,840
1060	2 post, semi-hydraulic, 10,000 lb	"	3,890	4,310	8,200
1070	Full hydraulic				
1080	10,000 lb	EA.	3,890	5,070	8,960
1100	13,000 lb	"	6,810	6,340	13,150
1120	18,500 lb	"	6,810	10,140	16,950
1140	24,000 lb	"	6,810	14,270	21,080
1160	26,000 lb	"	6,810	13,890	20,700
1170	Pneumatic hoist, fully hydraulic				
1180	11,000 lb	EA.	9,080	6,850	15,930
1200	24,000 lb	"	9,080	12,370	21,450
14560.10	**CHUTES**				
1020	Linen chutes, stainless steel, with supports				
1030	18" dia.	L.F.	4.97	160	165
1040	24" dia.	"	5.35	210	215
1050	30" dia.	"	5.80	220	226
1060	Hopper	EA.	46.50	2,660	2,707
1070	Skylight	"	70.00	1,620	1,690
14580.10	**PNEUMATIC SYSTEMS**				
6000	Air-lift conveyor, 115 Volt, single phase, 100' long, 3" carrier	EA.	5,830	4,420	10,250
6010	4" carrier	"	5,830	5,490	11,320
6020	6" carrier	"	5,830	9,970	15,800
7000	Pneumatic tube system accessories				
7010	Couplings and hanging accessories for 3" carrier system	L.F.			7.55
7020	For 4" carrier system	"			10.75
7030	For 6" carrier system	"			14.50
7040	24" CL. Expanded 90° bend, heavy duty, 3"-6" carrier systems	EA.			130
7050	36" CL. Expanded 90° bend, heavy duty, 3"-6" carrier systems	"			310
7060	48" CL. Expanded 45° bend, heavy duty, 3" carrier system	"			100
7070	4" carrier system	"			140
7080	6" carrier system	"			210
7090	48" CL. Expanded 90° bend, heavy duty, 3" carrier system	"			130
8000	4" carrier system	"			180
8010	6" carrier system	"			320
8020	Stainless steel body up-grade, 3" carrier system	"			440
8030	4" carrier system	"			450
8040	6" carrier system	"			640
14580.30	**DUMBWAITERS**				
0010	28' travel, extruded alum., 4 stops, 100 lbs. capacity	EA.			6,530
0020	150 lbs. capacity	"			9,190
0030	200 lbs. capacity	"			13,190

Design Cost Data™ **DCD**

TABLE OF CONTENTS   PAGE

		UNIT	LABOR	MAT.	TOTAL
15120.10	**BACKFLOW PREVENTERS**				
0080	Backflow preventer, flanged, cast iron, with valves				
0100	3" pipe	EA.	350	4,030	4,380
15410.06	**C.I. PIPE, BELOW GROUND**				
1010	No hub pipe				
1020	1-1/2" pipe	L.F.	3.45	8.66	12.11
1030	2" pipe	"	3.84	8.89	12.73
1120	3" pipe	"	4.32	12.25	16.57
1220	4" pipe	"	5.76	16.00	21.76
15410.10	**COPPER PIPE**				
0880	Type "K" copper				
0900	1/2"	L.F.	2.16	3.77	5.93
1000	3/4"	"	2.30	7.03	9.33
1020	1"	"	2.46	9.20	11.66
3000	DWV, copper				
3020	1-1/4"	L.F.	2.88	10.25	13.13
3030	1-1/2"	"	3.14	13.00	16.14
3040	2"	"	3.45	17.00	20.45
3070	3"	"	3.84	29.00	32.84
3080	4"	"	4.32	50.00	54.32
3090	6"	"	4.93	200	205
6080	Type "L" copper				
6090	1/4"	L.F.	2.03	1.52	3.55
6095	3/8"	"	2.03	2.33	4.36
6100	1/2"	"	2.16	2.71	4.87
6190	3/4"	"	2.30	4.33	6.63
6240	1"	"	2.46	6.50	8.96
6580	Type "M" copper				
6600	1/2"	L.F.	2.16	1.91	4.07
6620	3/4"	"	2.30	3.12	5.42
6630	1"	"	2.46	5.06	7.52
15410.11	**COPPER FITTINGS**				
0850	Slip coupling				
0860	1/4"	EA.	23.00	0.78	23.78
0870	1/2"	"	27.75	1.31	29.06
0880	3/4"	"	34.50	2.74	37.24
0890	1"	"	38.50	5.82	44.32
2660	Street ells, copper				
2670	1/4"	EA.	27.75	5.69	33.44
2680	3/8"	"	31.50	3.92	35.42
2690	1/2"	"	34.50	1.58	36.08
2700	3/4"	"	36.25	3.33	39.58
2710	1"	"	38.50	8.62	47.12
4190	DWV fittings, coupling with stop				
4210	1-1/2"	EA.	43.25	6.01	49.26
4230	2"	"	46.00	8.32	54.32
4260	3"	"	58.00	16.00	74.00
4280	3" x 2"	"	58.00	36.75	94.75
4290	4"	"	69.00	51.00	120
4300	Slip coupling				
4310	1-1/2"	EA.	43.25	9.34	52.59
4320	2"	"	46.00	11.00	57.00

		UNIT	LABOR	MAT.	TOTAL
15410.11	**COPPER FITTINGS, Cont'd...**				
4330	3"	EA.	58.00	20.25	78.25
4340	90 ells				
4350	1-1/2"	EA.	43.25	11.50	54.75
4360	1-1/2" x 1-1/4"	"	43.25	31.25	74.50
4370	2"	"	46.00	20.75	66.75
4380	2" x 1-1/2"	"	46.00	42.00	88.00
4390	3"	"	58.00	55.00	113
4400	4"	"	69.00	180	249
4410	Street, 90 elbows				
4420	1-1/2"	EA.	43.25	14.50	57.75
4430	2"	"	46.00	31.75	77.75
4440	3"	"	58.00	81.00	139
4450	4"	"	69.00	200	269
5410	No-hub adapters				
5420	1-1/2" x 2"	EA.	43.25	26.00	69.25
5430	2"	"	46.00	24.50	70.50
5440	2" x 3"	"	46.00	56.00	102
5450	3"	"	58.00	49.25	107
5460	3" x 4"	"	58.00	100	158
5470	4"	"	69.00	110	179
15410.82	**GALVANIZED STEEL PIPE**				
1000	Galvanized pipe				
1020	1/2" pipe	L.F.	6.91	3.01	9.92
1040	3/4" pipe	"	8.64	3.92	12.56
1200	90 degree ell, 150 lb malleable iron, galvanized				
1210	1/2"	EA.	13.75	2.02	15.77
1220	3/4"	"	17.25	2.68	19.93
1400	45 degree ell, 150 lb m.i., galv.				
1410	1/2"	EA.	13.75	3.23	16.98
1420	3/4"	"	17.25	4.38	21.63
1520	Tees, straight, 150 lb m.i., galv.				
1530	1/2"	EA.	17.25	2.68	19.93
1540	3/4"	"	19.75	4.47	24.22
1800	Couplings, straight, 150 lb m.i., galv.				
1810	1/2"	EA.	13.75	2.48	16.23
1820	3/4"	"	15.25	2.98	18.23
15430.23	**CLEANOUTS**				
0980	Cleanout, wall				
1000	2"	EA.	46.00	240	286
1020	3"	"	46.00	340	386
1040	4"	"	58.00	340	398
1050	Floor				
1060	2"	EA.	58.00	220	278
1080	3"	"	58.00	290	348
1100	4"	"	69.00	300	369
15430.25	**HOSE BIBBS**				
0005	Hose bibb				
0010	1/2"	EA.	23.00	10.50	33.50
0200	3/4"	"	23.00	11.00	34.00

		UNIT	LABOR	MAT.	TOTAL
15430.60	**VALVES**				
0780	Gate valve, 125 lb, bronze, soldered				
0800	1/2"	EA.	17.25	34.50	51.75
1000	3/4"	"	17.25	41.25	58.50
1280	Check valve, bronze, soldered, 125 lb				
1300	1/2"	EA.	17.25	59.00	76.25
1320	3/4"	"	17.25	74.00	91.25
1790	Globe valve, bronze, soldered, 125 lb				
1800	1/2"	EA.	19.75	73.00	92.75
1810	3/4"	"	21.50	90.00	112
15430.65	**VACUUM BREAKERS**				
1000	Vacuum breaker, atmospheric, threaded connection				
1010	3/4"	EA.	27.75	55.00	82.75
1018	Anti-siphon, brass				
1020	3/4"	EA.	27.75	60.00	87.75
15430.68	**STRAINERS**				
0980	Strainer, Y pattern, 125 psi, cast iron body, threaded				
1000	3/4"	EA.	24.75	15.25	40.00
1980	250 psi, brass body, threaded				
2000	3/4"	EA.	27.75	36.00	63.75
2130	Cast iron body, threaded				
2140	3/4"	EA.	27.75	19.00	46.75
15430.70	**DRAINS, ROOF & FLOOR**				
1020	Floor drain, cast iron, with cast iron top				
1030	2"	EA.	58.00	180	238
1040	3"	"	58.00	180	238
1050	4"	"	58.00	390	448
1090	Roof drain, cast iron				
1100	2"	EA.	58.00	280	338
1110	3"	"	58.00	290	348
1120	4"	"	58.00	370	428
15440.10	**BATHS**				
0980	Bath tub, 5' long				
1000	Minimum	EA.	230	580	810
1020	Average	"	350	1,270	1,620
1040	Maximum	"	690	2,900	3,590
1050	6' long				
1060	Minimum	EA.	230	650	880
1080	Average	"	350	1,330	1,680
1100	Maximum	"	690	3,760	4,450
1110	Square tub, whirlpool, 4'x4'				
1120	Minimum	EA.	350	2,000	2,350
1140	Average	"	690	2,830	3,520
1160	Maximum	"	860	8,640	9,500
1170	5'x5'				
1180	Minimum	EA.	350	2,000	2,350
1200	Average	"	690	2,830	3,520
1220	Maximum	"	860	8,800	9,660
1230	6'x6'				
1240	Minimum	EA.	350	2,430	2,780
1260	Average	"	690	3,560	4,250
1280	Maximum	"	860	10,200	11,060

		UNIT	LABOR	MAT.	TOTAL
15440.10	**BATHS, Cont'd...**				
8980	For trim and rough-in				
9000	Minimum	EA.	230	210	440
9020	Average	"	350	300	650
9040	Maximum	"	690	860	1,550
15440.12	**DISPOSALS & ACCESSORIES**				
0040	Disposal, continuous feed				
0050	Minimum	EA.	140	79.00	219
0060	Average	"	170	220	390
0070	Maximum	"	230	420	650
0200	Batch feed, 1/2 hp				
0220	Minimum	EA.	140	300	440
0230	Average	"	170	600	770
0240	Maximum	"	230	1,040	1,270
1100	Hot water dispenser				
1110	Minimum	EA.	140	220	360
1120	Average	"	170	350	520
1130	Maximum	"	230	560	790
1140	Epoxy finish faucet	"	140	310	450
1160	Lock stop assembly	"	86.00	67.00	153
1170	Mounting gasket	"	58.00	7.74	65.74
1180	Tailpipe gasket	"	58.00	1.13	59.13
1190	Stopper assembly	"	69.00	26.50	95.50
1200	Switch assembly, on/off	"	120	30.25	150
1210	Tailpipe gasket washer	"	34.50	1.21	35.71
1220	Stop gasket	"	38.50	2.66	41.16
1230	Tailpipe flange	"	34.50	0.30	34.80
1240	Tailpipe	"	43.25	3.44	46.69
15440.15	**FAUCETS**				
0980	Kitchen				
1000	Minimum	EA.	120	91.00	211
1020	Average	"	140	250	390
1040	Maximum	"	170	310	480
1050	Bath				
1060	Minimum	EA.	120	91.00	211
1080	Average	"	140	270	410
1100	Maximum	"	170	410	580
1110	Lavatory, domestic				
1120	Minimum	EA.	120	97.00	217
1140	Average	"	140	310	450
1160	Maximum	"	170	510	680
1290	Washroom				
1300	Minimum	EA.	120	120	240
1320	Average	"	140	300	440
1340	Maximum	"	170	560	730
1350	Handicapped				
1360	Minimum	EA.	140	130	270
1380	Average	"	170	400	570
1400	Maximum	"	230	620	850
1410	Shower				
1420	Minimum	EA.	120	120	240
1440	Average	"	140	350	490

		UNIT	LABOR	MAT.	TOTAL
15440.15	**FAUCETS, Cont'd...**				
1460	Maximum	EA.	170	560	730
1480	For trim and rough-in				
1500	Minimum	EA.	140	85.00	225
1520	Average	"	170	130	300
1540	Maximum	"	350	220	570
15440.18	**HYDRANTS**				
0980	Wall hydrant				
1000	8" thick	EA.	120	400	520
1020	12" thick	"	140	470	610
15440.20	**LAVATORIES**				
1980	Lavatory, counter top, porcelain enamel on cast iron				
2000	Minimum	EA.	140	210	350
2010	Average	"	170	320	490
2020	Maximum	"	230	570	800
2080	Wall hung, china				
2100	Minimum	EA.	140	290	430
2110	Average	"	170	340	510
2120	Maximum	"	230	850	1,080
2280	Handicapped				
2300	Minimum	EA.	170	470	640
2310	Average	"	230	540	770
2320	Maximum	"	350	910	1,260
8980	For trim and rough-in				
9000	Minimum	EA.	170	240	410
9020	Average	"	230	410	640
9040	Maximum	"	350	510	860
15440.30	**SHOWERS**				
0980	Shower, fiberglass, 36"x34"x84"				
1000	Minimum	EA.	490	630	1,120
1020	Average	"	690	880	1,570
1040	Maximum	"	690	1,270	1,960
2980	Steel, 1 piece, 36"x36"				
3000	Minimum	EA.	490	580	1,070
3020	Average	"	690	880	1,570
3040	Maximum	"	690	1,040	1,730
3980	Receptor, molded stone, 36"x36"				
4000	Minimum	EA.	230	240	470
4020	Average	"	350	410	760
4040	Maximum	"	580	630	1,210
8980	For trim and rough-in				
9000	Minimum	EA.	310	240	550
9020	Average	"	380	410	790
9040	Maximum	"	690	510	1,200
15440.40	**SINKS**				
0980	Service sink, 24"x29"				
1000	Minimum	EA.	170	700	870
1020	Average	"	230	870	1,100
1040	Maximum	"	350	1,280	1,630
2000	Kitchen sink, single, stainless steel, single bowl				
2020	Minimum	EA.	140	310	450
2040	Average	"	170	350	520

		UNIT	LABOR	MAT.	TOTAL
15440.40	**SINKS, Cont'd...**				
2060	Maximum	EA.	230	640	870
2070	Double bowl				
2080	Minimum	EA.	170	350	520
2100	Average	"	230	390	620
2120	Maximum	"	350	680	1,030
2190	Porcelain enamel, cast iron, single bowl				
2200	Minimum	EA.	140	220	360
2220	Average	"	170	290	460
2240	Maximum	"	230	450	680
2250	Double bowl				
2260	Minimum	EA.	170	300	470
2280	Average	"	230	420	650
2300	Maximum	"	350	600	950
2980	Mop sink, 24"x36"x10"				
3000	Minimum	EA.	140	530	670
3020	Average	"	170	640	810
3040	Maximum	"	230	860	1,090
5980	Washing machine box				
6000	Minimum	EA.	170	190	360
6040	Average	"	230	280	510
6060	Maximum	"	350	340	690
8980	For trim and rough-in				
9000	Minimum	EA.	230	310	540
9020	Average	"	350	480	830
9040	Maximum	"	460	620	1,080
15440.60	**WATER CLOSETS**				
0980	Water closet flush tank, floor mounted				
1000	Minimum	EA.	170	360	530
1010	Average	"	230	710	940
1020	Maximum	"	350	1,130	1,480
1030	Handicapped				
1040	Minimum	EA.	230	490	720
1050	Average	"	350	890	1,240
1060	Maximum	"	690	1,680	2,370
8980	For trim and rough-in				
9000	Minimum	EA.	170	230	400
9020	Average	"	230	270	500
9040	Maximum	"	350	360	710
15440.70	**DOMESTIC WATER HEATERS**				
0980	Water heater, electric				
1000	6 gal	EA.	120	450	570
1020	10 gal	"	120	460	580
1030	15 gal	"	120	450	570
1040	20 gal	"	140	630	770
1050	30 gal	"	140	650	790
1060	40 gal	"	140	710	850
1070	52 gal	"	170	800	970
2980	Oil fired				
3000	20 gal	EA.	350	1,430	1,780
3020	50 gal	"	490	2,230	2,720

		UNIT	LABOR	MAT.	TOTAL
15610.10	**FURNACES**				
0980	Electric, hot air				
1000	40 mbh	EA.	350	850	1,200
1020	60 mbh	"	360	920	1,280
1040	80 mbh	"	380	1,000	1,380
1060	100 mbh	"	410	1,130	1,540
1080	125 mbh	"	420	1,380	1,800
1980	Gas fired hot air				
2000	40 mbh	EA.	350	850	1,200
2020	60 mbh	"	360	910	1,270
2040	80 mbh	"	380	1,050	1,430
2060	100 mbh	"	410	1,090	1,500
2080	125 mbh	"	420	1,200	1,620
2980	Oil fired hot air				
3000	40 mbh	EA.	350	1,140	1,490
3020	60 mbh	"	360	1,890	2,250
3040	80 mbh	"	380	1,900	2,280
3060	100 mbh	"	410	1,930	2,340
3080	125 mbh	"	420	2,000	2,420
15780.20	**ROOFTOP UNITS**				
0980	Packaged, single zone rooftop unit, with roof curb				
1000	2 ton	EA.	690	3,930	4,620
1020	3 ton	"	690	4,130	4,820
1040	4 ton	"	860	4,510	5,370
15830.70	**UNIT HEATERS**				
0980	Steam unit heater, horizontal				
1000	12,500 btuh, 200 cfm	EA.	120	560	680
1010	17,000 btuh, 300 cfm	"	120	740	860
15855.10	**AIR HANDLING UNITS**				
0980	Air handling unit, medium pressure, single zone				
1000	1500 cfm	EA.	430	4,840	5,270
1060	3000 cfm	"	770	6,360	7,130
8980	Rooftop air handling units				
9000	4950 cfm	EA.	770	13,910	14,680
9060	7370 cfm	"	990	17,640	18,630
15870.20	**EXHAUST FANS**				
0160	Belt drive roof exhaust fans				
1020	640 cfm, 2618 fpm	EA.	86.00	1,090	1,176
1030	940 cfm, 2604 fpm	"	86.00	1,410	1,496
15890.10	**METAL DUCTWORK**				
0090	Rectangular duct				
0100	Galvanized steel				
1000	Minimum	Lb.	6.28	0.88	7.16
1010	Average	"	7.68	1.10	8.78
1020	Maximum	"	11.50	1.68	13.18
1080	Aluminum				
1100	Minimum	Lb.	13.75	2.29	16.04
1120	Average	"	17.25	3.05	20.30
1140	Maximum	"	23.00	3.79	26.79
1160	Fittings				
1180	Minimum	EA.	23.00	7.26	30.26
1200	Average	"	34.50	11.00	45.50

		UNIT	LABOR	MAT.	TOTAL
15890.10	**METAL DUCTWORK, Cont'd...**				
1220	Maximum	EA.	69.00	16.00	85.00
15890.30	**FLEXIBLE DUCTWORK**				
1010	Flexible duct, 1.25" fiberglass				
1020	5" dia.	L.F.	3.45	3.31	6.76
1040	6" dia.	"	3.84	3.68	7.52
1060	7" dia.	"	4.06	4.54	8.60
1080	8" dia.	"	4.32	4.76	9.08
1100	10" dia.	"	4.93	6.34	11.27
1120	12" dia.	"	5.31	6.93	12.24
9000	Flexible duct connector, 3" wide fabric	"	11.50	2.31	13.81
15910.10	**DAMPERS**				
0980	Horizontal parallel aluminum backdraft damper				
1000	12" x 12"	EA.	17.25	58.00	75.25
1010	16" x 16"	"	19.75	60.00	79.75
15940.10	**DIFFUSERS**				
1980	Ceiling diffusers, round, baked enamel finish				
2000	6" dia.	EA.	23.00	40.25	63.25
2020	8" dia.	"	28.75	48.50	77.25
2040	10" dia.	"	28.75	54.00	82.75
2060	12" dia.	"	28.75	69.00	97.75
2480	Rectangular				
2500	6x6"	EA.	23.00	43.00	66.00
2520	9x9"	"	34.50	52.00	86.50
2540	12x12"	"	34.50	76.00	111
2560	15x15"	"	34.50	95.00	130
2580	18x18"	"	34.50	120	155

DCD

Design Cost Data™

TABLE OF CONTENTS   PAGE

		UNIT	LABOR	MAT.	TOTAL
16050.30	**BUS DUCT**				
1000	Bus duct, 100a, plug-in				
1010	10', 600v	EA.	220	320	540
1020	With ground	"	340	430	770
1145	Circuit breakers, with enclosure				
1147	1 pole				
1150	15a-60a	EA.	80.00	320	400
1160	70a-100a	"	100	370	470
1165	2 pole				
1170	15a-60a	EA.	88.00	450	538
1180	70a-100a	"	100	540	640
16110.22	**EMT CONDUIT**				
0080	EMT conduit				
0100	1/2"	L.F.	2.42	0.60	3.02
1020	3/4"	"	3.20	1.09	4.29
1030	1"	"	4.01	1.82	5.83
2980	90 deg. elbow				
3000	1/2"	EA.	7.12	5.63	12.75
3040	3/4"	"	8.02	6.19	14.21
3060	1"	"	8.55	9.55	18.10
3980	Connector, steel compression				
4000	1/2"	EA.	7.12	1.72	8.84
4040	3/4"	"	7.12	3.30	10.42
4060	1"	"	7.12	4.97	12.09
0080	Flexible conduit, steel				
0100	3/8"	L.F.	2.42	0.75	3.17
1020	1/2	"	2.42	0.85	3.27
1040	3/4"	"	3.20	1.17	4.37
1060	1"	"	3.20	2.23	5.43
16110.24	**GALVANIZED CONDUIT**				
1980	Galvanized rigid steel conduit				
2000	1/2"	L.F.	3.20	2.82	6.02
2040	3/4"	"	4.01	3.13	7.14
2060	1"	"	4.75	4.51	9.26
2080	1-1/4"	"	6.41	6.24	12.65
2100	1-1/2"	"	7.12	7.34	14.46
2120	2"	"	8.02	9.34	17.36
2480	90 degree ell				
2500	1/2"	EA.	20.00	7.51	27.51
2540	3/4"	"	24.75	7.85	32.60
2560	1"	"	30.50	12.00	42.50
2580	1-1/4"	"	35.75	16.50	52.25
2590	1-1/2"	"	40.00	20.50	60.50
2600	2"	"	42.75	29.50	72.25
3200	Couplings, with set screws				
3220	1/2"	EA.	4.01	3.76	7.77
3260	3/4"	"	4.75	4.96	9.71
3280	1"	"	6.41	7.90	14.31
3300	1-1/4"	"	8.02	13.50	21.52
3320	1-1/2"	"	9.87	17.50	27.37
3340	2"	"	11.75	39.25	51.00

		UNIT	LABOR	MAT.	TOTAL
16110.25	**PLASTIC CONDUIT**				
3010	PVC conduit, schedule 40				
3020	1/2"	L.F.	2.42	0.73	3.15
3040	3/4"	"	2.42	0.91	3.33
3060	1"	"	3.20	1.32	4.52
3080	1-1/4"	"	3.20	1.82	5.02
3100	1-1/2"	"	4.01	2.17	6.18
3120	2"	"	4.01	2.78	6.79
3480	Couplings				
3500	1/2"	EA.	4.01	0.45	4.46
3520	3/4"	"	4.01	0.53	4.54
3540	1"	"	4.01	0.84	4.85
3560	1-1/4"	"	4.75	1.11	5.86
3580	1-1/2"	"	4.75	1.54	6.29
3600	2"	"	4.75	2.02	6.77
3705	90 degree elbows				
3710	1/2"	EA.	8.02	1.74	9.76
3740	3/4"	"	9.87	1.90	11.77
3760	1"	"	9.87	3.01	12.88
3780	1-1/4"	"	11.75	4.20	15.95
3800	1-1/2"	"	15.25	5.69	20.94
3810	2"	"	17.75	7.94	25.69
16110.28	**STEEL CONDUIT**				
7980	Intermediate metal conduit (IMC)				
8000	1/2"	L.F.	2.42	2.02	4.44
8040	3/4"	"	3.20	2.48	5.68
8060	1"	"	4.01	3.76	7.77
8080	1-1/4"	"	4.75	4.81	9.56
8100	1-1/2"	"	6.41	6.02	12.43
8120	2"	"	7.12	7.86	14.98
8490	90 degree ell				
8500	1/2"	EA.	20.00	15.50	35.50
8540	3/4"	"	24.75	16.25	41.00
8560	1"	"	30.50	24.75	55.25
8580	1-1/4"	"	35.75	34.50	70.25
8600	1-1/2"	"	40.00	42.50	82.50
8620	2"	"	45.75	61.00	107
9260	Couplings				
9280	1/2"	EA.	4.01	3.79	7.80
9290	3/4"	"	4.75	4.66	9.41
9300	1"	"	6.41	6.90	13.31
9310	1-1/4"	"	7.12	8.64	15.76
9320	1-1/2"	"	8.02	11.00	19.02
9330	2"	"	8.55	14.50	23.05
16110.35	**SURFACE MOUNTED RACEWAY**				
0980	Single Raceway				
1000	3/4" x 17/32" Conduit	L.F.	3.20	1.83	5.03
1020	Mounting Strap	EA.	4.27	0.49	4.76
1040	Connector	"	4.27	0.66	4.93
1060	Elbow				
2000	45 degree	EA.	4.01	8.38	12.39
2020	90 degree	"	4.01	2.67	6.68

		UNIT	LABOR	MAT.	TOTAL
16110.35	**SURFACE MOUNTED RACEWAY, Cont'd...**				
2040	internal	EA.	4.01	3.36	7.37
2050	external	"	4.01	3.10	7.11
2060	Switch	"	32.00	21.75	53.75
2100	Utility Box	"	32.00	14.50	46.50
2110	Receptacle	"	32.00	25.75	57.75
2140	3/4" x 21/32" Conduit	L.F.	3.20	2.09	5.29
2160	Mounting Strap	EA.	4.27	0.77	5.04
2180	Connector	"	4.27	0.79	5.06
2200	Elbow				
2210	45 degree	EA.	4.01	10.25	14.26
2220	90 degree	"	4.01	2.85	6.86
2240	internal	"	4.01	3.87	7.88
2260	external	"	4.01	3.87	7.88
3000	Switch	"	32.00	21.75	53.75
3010	Utility Box	"	32.00	14.50	46.50
3020	Receptacle	"	32.00	25.75	57.75
16120.43	**COPPER CONDUCTORS**				
0980	Copper conductors, type THW, solid				
1000	#14	L.F.	0.32	0.12	0.44
1040	#12	"	0.40	0.18	0.58
1060	#10	"	0.48	0.28	0.76
2010	THHN-THWN, solid				
2020	#14	L.F.	0.32	0.12	0.44
2040	#12	"	0.40	0.18	0.58
2060	#10	"	0.48	0.28	0.76
6215	Type "BX" solid armored cable				
6220	#14/2	L.F.	2.00	0.81	2.81
6230	#14/3	"	2.25	1.28	3.53
6240	#14/4	"	2.46	1.80	4.26
6250	#12/2	"	2.25	0.83	3.08
6260	#12/3	"	2.46	1.34	3.80
6270	#12/4	"	2.78	1.85	4.63
6280	#10/2	"	2.46	1.55	4.01
6290	#10/3	"	2.78	2.22	5.00
6300	#10/4	"	3.20	3.45	6.65
16120.47	**SHEATHED CABLE**				
6700	Non-metallic sheathed cable				
6705	Type NM cable with ground				
6710	#14/2	L.F.	1.19	0.28	1.47
6720	#12/2	"	1.28	0.44	1.72
6730	#10/2	"	1.42	0.69	2.11
6740	#8/2	"	1.60	1.13	2.73
6750	#6/2	"	2.00	1.78	3.78
6760	#14/3	"	2.06	0.39	2.45
6770	#12/3	"	2.13	0.62	2.75
6780	#10/3	"	2.17	0.99	3.16
6790	#8/3	"	2.21	1.66	3.87
6800	#6/3	"	2.25	2.68	4.93
6810	#4/3	"	2.56	5.55	8.11
6820	#2/3	"	2.78	8.34	11.12

		UNIT	LABOR	MAT.	TOTAL
16130.40	**BOXES**				
5000	Round cast box, type SEH				
5010	1/2"	EA.	28.00	20.75	48.75
5020	3/4"	"	33.75	20.75	54.50
16130.60	**PULL AND JUNCTION BOXES**				
1050	4"				
1060	Octagon box	EA.	9.16	3.77	12.93
1070	Box extension	"	4.75	6.35	11.10
1080	Plaster ring	"	4.75	3.48	8.23
1100	Cover blank	"	4.75	1.54	6.29
1120	Square box	"	9.16	5.43	14.59
1140	Box extension	"	4.75	5.32	10.07
1160	Plaster ring	"	4.75	2.91	7.66
1180	Cover blank	"	4.75	1.49	6.24
16130.80	**RECEPTACLES**				
0500	Contractor grade duplex receptacles, 15a 120v				
0510	Duplex	EA.	16.00	1.60	17.60
1000	125 volt, 20a, duplex, standard grade	"	16.00	12.00	28.00
1040	Ground fault interrupter type	"	23.75	38.75	62.50
1520	250 volt, 20a, 2 pole, single, ground type	"	16.00	20.00	36.00
1540	120/208v, 4 pole, single receptacle, twist lock				
1560	20a	EA.	28.00	23.75	51.75
1580	50a	"	28.00	45.25	73.25
1590	125/250v, 3 pole, flush receptacle				
1600	30a	EA.	23.75	24.00	47.75
1620	50a	"	23.75	29.75	53.50
1640	60a	"	28.00	77.00	105
16350.10	**CIRCUIT BREAKERS**				
5000	Load center circuit breakers, 240v				
5010	1 pole, 10-60a	EA.	20.00	20.25	40.25
5015	2 pole				
5020	10-60a	EA.	32.00	41.00	73.00
5030	70-100a	"	53.00	120	173
5040	110-150a	"	58.00	260	318
5065	Load center, G.F.I. breakers, 240v				
5070	1 pole, 15-30a	EA.	23.75	150	174
5095	Tandem breakers, 240v				
5100	1 pole, 15-30a	EA.	32.00	33.25	65.25
5110	2 pole, 15-30a	"	42.75	61.00	104
16395.10	**GROUNDING**				
0500	Ground rods, copper clad, 1/2" x				
0510	6'	EA.	53.00	16.25	69.25
0520	8'	"	58.00	22.25	80.25
0535	5/8" x				
0550	6'	EA.	58.00	21.50	79.50
0560	8'	"	80.00	28.00	108
16430.20	**METERING**				
0500	Outdoor wp meter sockets, 1 gang, 240v, 1 phase				
0510	Includes sealing ring, 100a	EA.	120	47.25	167
0520	150a	"	140	56.00	196
0530	200a	"	160	71.00	231

		UNIT	LABOR	MAT.	TOTAL
16470.10	**PANELBOARDS**				
1000	Indoor load center, 1 phase 240v main lug only				
1020	30a - 2 spaces	EA.	160	32.25	192
1030	100a - 8 spaces	"	190	100	290
1040	150a - 16 spaces	"	240	270	510
1050	200a - 24 spaces	"	280	550	830
1060	200a - 42 spaces	"	320	570	890
16490.10	**SWITCHES**				
4000	Photo electric switches				
4010	1000 watt				
4020	105-135v	EA.	58.00	37.00	95.00
4970	Dimmer switch and switch plate				
4990	600w	EA.	24.75	34.00	58.75
5171	Contractor grade wall switch 15a, 120v				
5172	Single pole	EA.	12.75	1.79	14.54
5173	Three way	"	16.00	3.26	19.26
5174	Four way	"	21.50	11.00	32.50
16510.05	**INTERIOR LIGHTING**				
0005	Recessed fluorescent fixtures, 2'x2'				
0010	2 lamp	EA.	58.00	72.00	130
0020	4 lamp	"	58.00	98.00	156
0205	Surface mounted incandescent fixtures				
0210	40w	EA.	53.00	110	163
0220	75w	"	53.00	110	163
0230	100w	"	53.00	120	173
0240	150w	"	53.00	160	213
0289	Recessed incandescent fixtures				
0290	40w	EA.	120	150	270
0300	75w	"	120	160	280
0310	100w	"	120	180	300
0320	150w	"	120	190	310
0395	Light track single circuit				
0400	2'	EA.	40.00	44.25	84.25
0410	4'	"	40.00	52.00	92.00
0420	8'	"	80.00	72.00	152
0430	12'	"	120	100	220
16510.10	**INDUSTRIAL LIGHTING**				
0500	Strip fluorescent				
0510	4'				
0520	1 lamp	EA.	53.00	45.50	98.50
0540	2 lamps	"	53.00	55.00	108
0550	8'				
0560	1 lamp	EA.	58.00	66.00	124
0580	2 lamps	"	71.00	99.00	170
1000	Parabolic troffer, 2'x2'				
1020	With 2 "U" lamps	EA.	80.00	130	210
1060	With 3 "U" lamps	"	92.00	150	242
1080	2'x4'				
1100	With 2 40w lamps	EA.	92.00	150	242
1120	With 3 40w lamps	"	110	150	260
1140	With 4 40w lamps	"	110	160	270
3120	High pressure sodium, hi-bay open				

		UNIT	LABOR	MAT.	TOTAL
16510.10	**INDUSTRIAL LIGHTING, Cont'd...**				
3140	400w	EA.	140	460	600
3160	1000w	"	190	790	980
3170	Enclosed				
3180	400w	EA.	190	740	930
3200	1000w	"	240	1,030	1,270
3210	Metal halide hi-bay, open				
3220	400w	EA.	140	280	420
3240	1000w	"	190	580	770
3250	Enclosed				
3260	400w	EA.	190	640	830
3280	1000w	"	240	610	850
3590	Metal halide, low bay, pendant mounted				
3600	175w	EA.	110	370	480
3620	250w	"	130	510	640
3660	400w	"	180	550	730
16710.10	**COMMUNICATIONS COMPONENTS**				
0020	Port desktop switch unit-4	EA.			85.00
0030	(USB)-4	"			220
0040	8	"			360
0050	16	"			600
0060	32	"			3,870
0070	Port console unit-8	"			2,420
0080	16	"			3,030
0090	Cat 5EJack and RJ45 coupler	"			4.11
1000	Quick-Port and voice grade	"			4.65
1010	Trac-Jack category 5E and connectors	"			4.63
1020	Snap-In connector	"			6.35
1030	Fast ethernet media converter	"			150
1040	Gigabit media converter	"			420
1050	With link fault signaling	"			150
1060	Gigabit switching media converter	"			600
1070	16-Bay media chassis	"			740
1080	Fiber-optic cable, Single-Mode, 50/125 microns, 10' length	"			24.25
1090	Multi-Mode, 62.5/125 microns, 10' length	"			27.75
2000	Fiber-Optic connectors, Unicam	"			18.75
2010	Fast-cam	"			19.25
2020	Adhesive style	"			6.95
2030	Threat-lock	"			12.00
2040	Unicam, high performance, single-mode	"			24.25
2050	multi-mode	"			21.75
2060	Network Cable, Cat5, solid PVC, 50'	"			17.50
2070	1000' Cat5, stranded PVC	"			200
2080	plenum PVC	"			240
2090	1000' Cat6, solid PVC	"			170
3000	USB Cables, 5 in 1 connector, male and female	"			21.75
3010	3 in 1 quick-connect, 4-pin or 6-pin	"			26.50
3020	Squid hub	"			36.25
16760.10	**AUDIO/VIDEO COMPONENTS**				
0010	Key-stone jack module	EA.			3.30
0020	F-Type quick port	"			2.31
0030	BNC quick port	"			5.50

		UNIT	LABOR	MAT.	TOTAL
16760.10	**AUDIO/VIDEO COMPONENTS, Cont'd...**				
0040	RCA jack bulkhead connector	EA.			4.40
0050	Quick-Port	"			5.72
0060	Snap-in	"			8.80
0070	HDMI wall plate and jack	"			10.50
0080	Voice/Data adapter	"			3.85
16760.20	**AUDIO/VIDEO CABLES**				
0010	Monster cable, 24k. gold plated	EA.			26.00
0020	DVI-HDMI	"			11.25
0030	Satellite/Video	"			3.76
0040	Digital/Optical	"			12.00
0050	Video/RCA	"			7.05
0060	S-Video	"			7.84
0070	Monster/S-video	"			41.75
0080	Speaker, clear jacket	FT.			0.89
0090	Speaker, high performance	"			2.09
0100	Flex-Premiere, oxygen free	"			0.31
16850.10	**ELECTRIC HEATING**				
1000	Baseboard heater				
1020	2', 375w	EA.	80.00	41.75	122
1040	3', 500w	"	80.00	49.50	130
1060	4', 750w	"	92.00	55.00	147
1100	5', 935w	"	110	78.00	188
1120	6', 1125w	"	130	92.00	222
1140	7', 1310w	"	150	100	250
1160	8', 1500w	"	160	120	280
1180	9', 1680w	"	180	130	310
1200	10', 1875w	"	180	180	360
1210	Unit heater, wall mounted				
1215	750w	EA.	130	160	290
1220	1500w	"	130	220	350
16910.40	**CONTROL CABLE**				
0980	Control cable, 600v, #14 THWN, PVC jacket				
1000	2 wire	L.F.	0.64	0.35	0.99
1020	4 wire	"	0.80	0.59	1.39

Square Foot Tables Explained

The following Square Foot Tables list hundreds of actual projects for dozens of building types, each with associated building size, total square foot building cost and percentage of project costs for total mechanical and electrical components. This data provides an overview of construction costs by building type. These costs are for actual projects. The variations within similar building types may be due, among other factors, to size, location, quality and specified components, materials and processes. Depending upon all such factors, specific building costs can vary significantly and may not necessarily fall within the range of costs as presented. The data has been updated to reflect current construction costs.

BUILDING CATEGORY PAGE

All prices are updated to January 1, 2019 and are national averages.
For a more in-depth report of any of these buildings or additional case studies contact
Design, Cost and Data at 800-533-5680, or go to www.DCD.com

SF-1

2019 SQUARE FOOT TABLES

PROJECT	DESCRIPTION	CITY	STATE	SIZE	$/SF	NOTES
		Commercial				
Bank	School Credit Union Administration Building	Katy	TX	30,700	$207.34	New
	FineMark National Bank & Trust	Fort Myers	FL	20,039	$391.17	New
	Florida Shores Bank	Pompano Beach	FL	11,697	$497.86	New
	Mobiloil Credit Union	Vidor	TX	9,252	$349.16	New
	Beaumont Community Credit Union	Beaumont	TX	3,267	$395.82	New
Office	Allendale Town Center	Allendale	NJ	80,226	$36.54	Addition/Renovation
	Roanoke Electric Cooperative	Ahoskie	NC	52,752	$208.43	New
	Regional Aviation & Training Center	Currituck	NC	39,930	$264.41	New
	Transportation/Warehouse Facility	Monroe	GA	32,400	$183.55	New
	Collection System Operations Facility	Walnut Creek	CA	27,179	$362.61	New
	ULTA - (Shell Only)	Pensacola	FL	10,850	$93.82	Renovation
	Daycare Center	Clawson	MI	4,270	$148.56	Adaptive Reuse
	Campgrounds Office & Retail	Cincinnati	OH	2,300	$325.46	New
Parking	Palm Avenue Parking Garage	Sarasota	FL	287,040	$55.60	New
	Reynolds Street Parking Deck	Augusta	GA	214,000	$66.14	New
	Awty Int. School Parking Structure	Houston	TX	174,582	$58.56	New
Retail	SNG Center - Mixed-Use	Fargo	ND	143,860	$98.41	New
	Roof & Lifeway/Steinmart Renovation	Pensacola	FL	88,299	$34.57	Renovation
	No Frills Supermarket	Omaha	NE	61,000	$105.28	New
	West Oaks Mall Redevelopment	Houston	TX	49,800	$187.36	Renovation
	World of Decor	Deerfield Beach	FL	47,500	$143.00	New
	Sarasota Yacht Club	Sarasota	FL	41,332	$420.17	New
	Karschs Village Market	Barnhart,	MO	35,384	$82.07	New
	Fresh Thyme Farmers Market	Fishers	IN	28,784	$179.94	New
	Nashville Hangar Inc.	Nashville	TN	28,702	$188.68	New
	Party Time Plus	Billings	MT	26,000	$95.73	New
	Marshalls - (Shell Only)	Pensacola	FL	25,990	$45.80	Renovation
	Ed Hicks Mercedes-Benz USA	Corpus Christi	TX	25,273	$250.38	New
	Montana Honda & Marine	Billings	MT	22,963	$108.80	Addition
	Fresh Market - (Shell Only)	Pensacola	FL	21,000	$78.74	Renovation
	Theatre Exchange Interior Fit Up	Manitoba	CA	19,344	$127.00	Renovation
	DSW Shoes Renovation	Pensacola	FL	18,000	$88.16	Renovation
	Dormans Lighting & Design	Lutherville	MD	15,220	$119.50	Addition
	The Groves Exterior Renovation	Farmington	MI	15,137	$67.56	Renovation
	Don Gibson Theatre	Shelby	NC	13,386	$356.20	Renovation
	Tri Ford Showroom Expansion	Highland	IL	12,881	$116.50	Addition/Renovation
	Fiat of LeHigh Valley	Easton	PA	11,905	$149.19	New
	Sicardi Art Gallery	Houston	TX	6,175	$234.68	New
	Childrens Mercy Hospital Gift Shop	Kansas City	MO	5,010	$295.77	Tenant Build-out
Restaurant	LaMar Cebicheria Peruana Restaurant	San Francisco	CA	11,000	$302.40	Tenant Build-out
	Ulele Restaurant	Tampa	FL	8,905	$732.93	Adaptive Reuse
	Youells Oyster House	Allentown	PA	6,107	$240.60	New
	Mellow Mushroom Highlands Shell	Louisville	KY	5,802	$97.52	New
	Mellow Mushroom Highlands TBO	Louisville	KY	5,802	$161.75	Tenant Build-out
	Mellow Mushroom Pizza	Wilder	KY	5,500	$231.34	New
	Liberty Microbrewery	Plymouth	MI	3,425	$166.45	Addition
	700 South Deli	Linthicum	MD	3,200	$205.09	Tenant Build-out
	New York Pizza Department (NYPD)	Tempe	AZ	2,338	$313.77	Tenant Build-out
	Airport Restaurant Build Out	Eglin Air Force Base	FL	2,320	$270.12	Tenant Build-out

All prices are updated to January 1, 2019 and are national averages.
For a more in-depth report of any of these buildings or additional case studies contact
Design, Cost and Data at 800-533-5680, or go to www.DCD.com

SF-3

2019 SQUARE FOOT TABLES

PROJECT	DESCRIPTION	CITY	STATE	SIZE	$/SF	NOTES
	Civic/Government					
Civic Center	Lincoln Center	Fort Collins	CO	38,160	206.87	Addition/Renovation
	Rockport City Services Building	Rockport	TX	20,062	214.01	New
	Sinclair Park Community Centre	Manitoba	CA	17,007	303.02	Addition/Renovation
	Mt. Olive City Hall Complex	Mount Olive	IL	14,360	116.20	New
	Teaneck Municipal Complex	Teaneck	NJ	12,870	222.68	Addition/Renovation
	Cobb Community Center Additions	Pensacola	FL	5,200	309.91	Addition/Renovation
	Newtown Municipal Center	Newtown	OH	5,077	161.25	Adaptive Reuse
	Nederland City Hall	Nederland	TX	4,983	358.76	New
Correctional	County Sheriffs Office	Morgantown	WV	31,645	284.34	New
	Nederland Public Safety Complex	Nederland	TX	21,189	207.84	Adaptive Reuse
	Detention Center & Sheriffs Office	Spencer	IA	16,983	403.07	New
	Ogle City Sheriff & Coroner Admin	Oregon	IL	15,377	271.59	New
	Chautauqua City Jail & Sheriff	Sedan	KS	12,257	286.44	New
Courthouse	Courthouse HVAC System Replacement	Gainesville	FL	101,000	41.31	Renovation
	Courthouse Renovation & Restoration	Springfield	IL	47,720	176.60	Renovation
Fire Department	College Station Fire Station #6	College Station	TX	25,133	334.92	New
	Mt. Orab Fire Station	Village of Mt. Orab	OH	18,170	172.81	New
	Richardson Fire Station No. 4	Richardson	TX	14,090	393.23	New
	Willowfork Fire Station No. 2	Katy	TX	13,358	289.04	New
	Joint Fire & Rescue Station	Newtown	OH	13,125	186.31	Addition/Renovation
	Little Miami Fire & Rescue	Fairfax	OH	12,316	221.24	New
	Fire Station No. 11	Fort Smith	AR	12,155	335.11	New
	Pearisburg Fire Station	Pearisburg	VA	11,818	215.69	New
	Ponderosa Fire Station No. 62	Spring	TX	11,163	289.37	New
	El Dorado Hills Fire Station 84	El Dorado Hills	CA	10,869	435.73	New
	Wayne Fire Department	Goshen	OH	10,000	94.84	New
	Fire Station No. 40	Jacksonville	FL	9,703	382.35	New
	Rosenberg Fire Station No. 3	Rosenberg	TX	8,479	362.45	New
	Little Rock Fire Station No. 23	Little Rock	AR	8,291	466.31	New
	Cleveland Volunteer Fire Station	Cleveland	MS	6,910	325.98	New
Government	Council Center For Scouting	Fargo	ND	20,466	195.50	New
	Camp Crook Ranger Station	Camp Crook	SD	4,880	482.86	New
	Beaumont Municipal Tennis Center	Beaumont	TX	4,460	353.31	Addition
	Florence Transit Hub	Florence	KY	3,115	504.68	New
	Knox Area Rescue Ministries	Knoxville	TN	1,762	664.66	New
	Entrance Station Lake Mead	Clark County	NV	480	2,696.62	New
	Vehicle Charging Stations	Denton	TX	6 spaces	6,918.73	New
Library	Dover Public Library	Dover	DE	46,424	423.90	New
	Clinton-Macomb Public Library	Clinton Township	MI	24,723	123.85	Adaptive Reuse
	Crozet Western Albemarle Library	Crozet	VA	23,199	344.14	New
	Upper Tampa Bay Regional Library	Tampa	FL	13,630	238.19	Addition/Renovation
	Palmetto Branch Library	Palmetto	GA	11,200	475.34	New
	Regional Library Expansion	Valrico	FL	10,970	291.98	Addition/Renovation
Miscellaneous	City of Pampa Animal Welfare	Pampa	TX	13,578	276.61	New
	Royal Winnipeg Ballet Renovations	Manitoba	CA	13,237	74.63	Renovation
	Senior Services - Kitchen Facility	Batavia	OH	6,000	118.13	New
	Historical Site Locomotive Shelter	Bismarck	ND	2,100	157.87	New
Office	Federal Building & Courthouse Modernization	Denver	CO	41,600	342.32	Renovation
	Brazos County Tax Office	Bryan	TX	13,143	294.53	New
	Illinois Water District Office	Lincoln	IL	8,974	153.28	New
	Arkansas River Resource Center	Little Rock	AR	4,926	562.79	New

All prices are updated to January 1, 2019 and are national averages.
For a more in-depth report of any of these buildings or additional case studies contact
Design, Cost and Data at 800-533-5680, or go to www.DCD.com

2019 SQUARE FOOT TABLES

PROJECT	DESCRIPTION	CITY	STATE	SIZE	$/SF	NOTES
	Educational					
Athletic Facility	Physical Activity/Sports Science	Morgantown	WV	117,344	233.70	New
	Indoor Football Practice Facility	Clemson	SC	81,992	180.33	New
	Jesuit College Locker Room Addition	Dallas	TX	41,673	215.61	Addition
	Multi-Purpose Gymnasium	Jacksonville	FL	22,844	286.11	New
College Classroom	New Mexico Tech Geology Building	Socorro	NM	86,813	311.68	New
	CSU Concourse & Training Room	Fort Collins	CO	53,050	145.59	Addition/Renovation
	Jack Williamson Liberal Arts Center	Portales	NM	52,480	233.93	Renovation
	UNLV Literature & Law Building	Las Vegas	NV	44,830	194.50	Renovation
	Southern State Community College	Mount Orab	OH	43,833	187.14	New
	SERT Building Iowa Lakes College	Estherville	IA	42,940	128.25	Adaptive Reuse
Elementary	Cibolo Valley Elementary School	Cibolo	TX	153,130	253.33	New
	Hill Farm Elementary School	Bryant	AR	88,800	303.03	New
	CREC International Magnet School	South Windsor	CT	63,923	416.51	New
	Pineville Elementary School	Pineville	WV	51,650	221.79	New
	Crownpoint Elementary School	Crownpoint	NM	48,592	413.26	New
	Janney Elementary School Addition	Washington	DC	10,000	610.81	Addition
High School	Hmong College Prep Academy	Saint Paul	MN	154,434	86.74	Addition/Renovation
	HFC High School North Building	Flossmoor	IL	136,555	212.30	Addition/Renovation
	Somerset Jr. High School	Von Ormy	TX	125,800	184.39	New
	Takoma Education Campus	Washington	DC	119,000	228.05	Renovation
	High School Addition & Renovation	Decatur	GA	83,816	264.78	Addition/Renovation
	Palmer Catholic Academy	Ponte Vedra Beach	FL	34,209	143.47	New
	Elmwood High School Addition	Elmwood Park	IL	31,630	355.25	Addition/Renovation
	High School Fine Arts Building	Heber Springs	AR	30,505	414.92	New
	Alamo Heights High School Fine Arts	San Antonio	TX	25,536	333.43	Addition/Renovation
	St. Patrick Catholic School	Jacksonville	FL	23,227	306.04	New
	Springdale School Alteration	Corbett	OR	13,680	130.57	Renovation
	Goddard School Addition Renovation	Anderson Township	OH	12,489	87.00	Addition/Renovation
	ISD Outdoor Education Center	Sabine Pass	TX	10,193	330.64	New
	Indian Mountain School Student Center	Lakeville	CT	9,335	317.42	Addition
	First Impressions Academy	Fayetteville	NC	7,752	182.63	New
	High School South Campus Field	Cincinnati	OH	5,580	244.93	New
Middle School	Timberline Middle School	Waukee	IA	187,375	164.40	New
	Red Bank Middle School	Chattanooga	TN	158,637	272.20	New
	Jaime Escalante Middle School	Pharr	TX	156,538	203.81	New
	Conservatory Green ECE-8 School	Denver	CO	113,616	181.76	New
	Midland Elementary School Addition	Floral	AR	30,150	244.42	Addition
Laboratory/Research	Science & Technology Building	Fayetteville	NC	65,048	515.66	New
	NSU/US Geological Survey	Davie	FL	24,000	107.16	Tenant Build-out
	Northeast Technology Center	Pryor	OK	11,909	321.43	New
	Environmental Education Center	Bushkill Township	PA	9,275	557.28	New
	Research & Education Center	Homestead	FL	5,760	703.26	New
Multi-Purpose	BGSU Student Rec Center	Bowling Green	OH	179,549	67.55	Renovation
	Kennedy Center Theatre/Studio Arts	Clinton	NY	96,100	341.13	New
	Classroom & Administration Building	Houston	TX	65,234	220.52	New
	NM State U Pete Domenici Building	Las Cruces	NM	53,341	277.95	Addition/Renovation
	Widener University Freedom Hall	Chester	PA	36,700	329.01	New
	Alumni Hall, Lincoln Park	Midland	PA	29,027	268.30	New
	GSU Piedmont North Dining Hall	Atlanta	GA	12,300	376.69	Addition
	NAU Dining Hall Expansion Phase II	Flagstaff	AZ	10,096	480.40	New
	Neighborhood Resource Center	Richmond	TX	6,935	207.38	New
	Heber Springs Cafeteria Remodel	Heber Springs	AR	5,585	347.51	Renovation

All prices are updated to January 1, 2019 and are national averages.
For a more in-depth report of any of these buildings or additional case studies contact
Design, Cost and Data at 800-533-5680, or go to www.DCD.com

SF-5

2019 SQUARE FOOT TABLES

PROJECT	DESCRIPTION	CITY	STATE	SIZE	$/SF	NOTES
	Hotels					
Hotels	Omni Dallas Hotel	Dallas	TX	1,161,450	424.50	New
	John Ascuagas Nugget Hotel/Casino	Sparks	NV	449,820	144.82	Addition
	Le Centre On Fourth Embassy Suites	Louisville	KY	408,229	114.61	Adaptive Reuse
	Minneapolis Marriott West	Minneapolis	MN	237,362	165.42	New
	Sheraton Centre Park Hotel	Dallas	TX	231,031	236.74	New
	AmeriSuites	Chicago	IL	191,600	149.69	Addition/Renovation
	Hampton Inn & Suites Hotel	Chicago	IL	162,000	163.47	New
	Sheraton Harbor Island Hotel Tower	San Diego	CA	144,126	199.30	Addition
	Compri Hotel	Los Angeles	CA	110,150	176.51	New
	Best Western Columbia Hotel	San Diego	CA	108,040	136.55	New
	Spooky Nook Warehouse Hotel	Manheim	PA	92,726	130.03	Adaptive Reuse
	The Atrium Motel	Norfolk	VA	75,889	139.64	New
	Staybridge Hotel At Preston Ridge	Alpharetta	GA	74,607	179.04	New
	Hampton Inn & Suites	Allentown	PA	71,686	149.32	New
	Hampton Inn Hotel	Carol Stream	IL	71,000	183.37	New
	The Inn On Lake Superior	Duluth	MN	65,345	147.51	New
	The Lancaster Hotel	Houston	TX	64,310	308.69	Renovation
	Fairfield Inn	Helena	MT	31,009	144.15	New
	The Edison Hotel	Miami	FL	28,875	119.50	Renovation
	Western Executive Inn	Billings	MT	21,984	106.70	New
	Country Hearth Inn	Preston	MN	21,028	125.27	New
	Lawrence Welk Resort Hotel	Escondido	CA	19,874	142.89	Addition
	Hanalei Hotel Conference Center	San Diego	CA	8,587	226.89	Addition
	Summit At Vail, Multi-Purpose Lodge	Vail	CO	6,000	270.90	New

2019 SQUARE FOOT TABLES

PROJECT	DESCRIPTION	CITY	STATE	SIZE	$/SF	NOTES

Industrial

PROJECT	DESCRIPTION	CITY	STATE	SIZE	$/SF	NOTES
Manufacturing	Brentwood Industries Manufacturing	Reading	PA	205,000	42.07	New
	Lee Steel Corporate Plant	Romulus	MI	200,625	91.50	New
	Siemens Westinghouse Fuel Cell Facility	Munhall	PA	191,090	99.63	New
	Manufacturing Plant & Headquarters	Lansing	MI	188,975	96.60	Addition
	Concepts Direct	Longmont	CO	117,900	117.82	New
	Nypro Inc.	Clinton	MA	102,475	120.86	Addition
	SWF Industrial	Wrightsville	PA	76,218	74.46	New
	Headquarters & Manufacturing Facility	Lower Nazareth Township	PA	62,980	35.67	Renovation
	Aerzen USA (Office/Manufacturing)	Coatesville	PA	40,000	181.99	New
	Lee Steel Corporate Expansion	Wyoming	MI	34,821	77.31	New
	Prescott Aerospace	Prescott	AZ	31,400	111.38	New
	Battery Innovation Center	Newberry	IN	30,080	434.58	New
	American Steel	Billings	MT	25,957	80.78	New
	Phillip S. Luttazi Town Garage	Dover	MA	21,913	139.01	New
	ITT Flygt - Industrial Facility	Milford	OH	16,991	131.81	New
	Broadmoor Golf Maintenance	Colorado Springs	CO	16,064	238.95	New
	Cooper B-Line Expansion	Highland	IL	15,290	168.82	Addition/Renovation
	Robberson Ford Collision Center	Bend	OR	15,089	138.59	New
	Brown Industrial Building	Truckee	CA	13,345	136.90	New
	CTC Vehicle Maintenance Shops	Killeen	TX	11,250	129.45	New
	Storage & Shop Facility	Billings	MT	8,763	89.33	New
	Central Plant with Equipment Bay	Mesa	AZ	8,500	921.08	New
Office	Woodlands Business Center	Richmond	VA	48,000	80.02	New
	Wiregrass Research Center	Headland	AL	9,740	247.13	New
Office/Warehouse	American Superconductor	Devens	MA	354,000	170.05	New
	Castcon Stone Inc.	Saxonburg	PA	47,000	113.90	New
	Minnesota DNR Headquarters	Tower	MN	37,802	168.68	New
	Office & Warehouse	Miami	FL	14,815	138.20	New
	DOT Office & Maintenance Building	Hillsboro	OH	10,876	243.77	New
Warehouse	Distribution Center	Windsor	CT	303,750	34.68	New
	Zany Brainy Distribution Center	Bridgeport	NJ	250,000	41.40	New
	Galderma - Warehouse	Fort Worth	TX	70,000	89.21	New
	Manzana Products Warehouse	Sebastopol	CA	41,395	66.44	New
	Tactical Equip Maintenance Facility	Fort Campbell	KY	35,290	239.71	New
	Sonoma Wine Company Canopy	Graton	CA	26,000	56.39	New
	Administration/Chemical Storage Building	Killeen	TX	23,837	340.98	New
	DOT Truck Storage Building	Hillsboro	OH	18,400	93.03	New
	F.I. Storage Facility	Kentwood	MI	13,125	68.41	New
	50 Columbia Drive Warehouse	Pooler	GA	10,000	85.85	New
	DOT Salt Storage Building	Hillsboro	OH	9,100	93.79	New
	Organizational Storage Facility	Fort Campbell	KY	8,040	142.31	New
	Dwan Maintenance Building	Bloomington	MN	7,240	192.20	Addition/Renovation
	Maintenance/Storage Building	Batavia	OH	7,200	82.67	New
	DOT Cold Storage Building	Hillsboro	OH	5,040	100.01	New
	Job Corp Warehouse	Hartford	CT	3,800	474.14	New
	DOT Materials Storage Building	Hillsboro	OH	1,920	111.81	New
	Aerial Vehicle Storage	Fort Campbell	KY	1,800	213.69	New
	Petro, Oil, Lubricant Storage	Fort Campbell	KY	640	298.72	New
	Hazardous Waste Storage Building	Fort Campbell	KY	640	319.24	New

All prices are updated to January 1, 2019 and are national averages.
For a more in-depth report of any of these buildings or additional case studies contact
Design, Cost and Data at 800-533-5680, or go to www.DCD.com

SF-7

2019 SQUARE FOOT TABLES

PROJECT	DESCRIPTION	CITY	STATE	SIZE	$/SF	NOTES
		Medical				
Clinic	HealthCare Emergency/Trauma Center	Topeka	KS	115,000	402.89	Addition
	Sadler Clinic (Shell Only)	Conroe	TX	61,599	125.42	New
	Pinellas County Health Department	Largo	FL	54,965	261.36	Retrofit
	Sanford Moorhead Clinic	Moorhead	MN	49,250	256.14	New
	County Health Department	Port Charlotte	FL	47,564	290.41	New
	Sadler Clinic	Conroe	TX	41,066	118.82	Tenant Build-out
	PineMed Medical Plaza	The Woodlands	TX	30,398	132.51	New
	Outpatient Specialty Clinic	Vancouver	WA	20,139	329.70	New
	Ambulatory Surgery Center	Stroudsburg	PA	19,929	353.85	Addition/Renovation
	Thundermist Health Center	West Warwick	RI	18,217	204.03	Adaptive Reuse
	E Texas Community Health Services	Nacogdoches	TX	12,500	96.63	Retrofit
	North Mobile Health Center	Mt. Vernon	AL	6,765	288.24	New
Dental Office	Construct Dental Clinic Roseburg	Roseburg	OR	7,750	501.57	New
	Kitchens Pediatric Dental Clinic	Little Rock	AR	6,068	301.93	New
	Dental Office Shell & Parking	Olympia	WA	5,302	171.31	New
	Evans Family Dental	Austin	TX	2,354	209.96	Tenant Build-out
Hospital	Union Hospital Addition	Terre Haute	IN	492,348	315.56	Addition
	Regional Medical Center	Lafayette	LA	410,273	528.13	New
	Houston Medical Pavilion	Warner Robins	GA	180,000	62.22	Adaptive Reuse
	Langley AFB Hospital Renovation	Langley Air Force Base	VA	160,000	527.26	Renovation
	Cass Regional Medical Center	Harrisonville	MO	137,524	370.75	New
	Texas Spine & Joint Hospital	Tyler	TX	115,789	245.56	Addition/Renovation
	Oktibbeha County Hospital Expansion	Starkville	MS	87,116	321.76	New
	Childrens Mercy Hospital	Independence	MO	54,682	310.01	New
	Oktibbeha County Hospital Renovation	Starkville	MS	30,263	160.60	Renovation
	El Rio Community Health Center	Tucson	AZ	26,998	236.78	New
	Pondella Public Health Center	Fort Myers	FL	26,400	294.93	Renovation
	Topeka Ear Nose & Throat	Topeka	KS	24,073	305.78	New
	Rapha Primary Care	Fayetteville	NC	19,907	131.27	Renovation
	UNM Hospitals North Valley Center	Albuquerque	NM	16,500	309.19	New
	El Rio Community Health Center	Tucson	AZ	14,000	268.82	New
	VA Medical Center Area G Renovation	Houston	TX	12,000	327.11	Renovation
	Surgical Suite Expansion	Dobbs Ferry	NY	9,000	373.35	Renovation
	Legacy Emergency Room	Allen	TX	8,432	619.24	New
	Oral & Maxillofacial Surgery Center	Fayetteville	NC	2,214	526.88	Renovation
Nursing Home/Rehab	Senior Living Community	Hoschton	GA	56,251	137.31	New
	Retirement Community	Carlisle	PA	47,075	154.63	Addition/Renovation
	Assisted Living & Memory Center	Dacula	GA	38,221	159.72	New
	Homestead Village Nursing Care	Lancaster	PA	28,149	104.57	Renovation
	Jewish Services For The Aging	Tucson	AZ	24,993	194.69	New
	Short-Term Rehabilitation	Olathe	KS	13,800	261.89	Addition
	St. Katharine Retirement Center	El Reno	OK	12,000	309.52	New/Addition
Office	Orthopedic Hospital/Medical Office	Allentown	PA	79,807	196.18	Renovation
	Tomball Medical Office Building	Tomball	TX	54,380	133.39	New
	Olathe Health Education Center	Olathe	KS	50,258	298.28	New
	Home & Hospice Care	Providence	RI	47,734	188.16	Renovation
	NE Georgia Medical Plaza 400	Dawsonville	GA	26,997	262.53	Adaptive Reuse
	VA Medical Center/Pharmacy	Waco	TX	19,171	152.85	Renovation
	Cancer Specialists of North Florida	Jacksonville	FL	18,654	278.82	New
	MJHS Hospice Residence	N.Y.C.	NY	12,500	144.27	Renovation
	Medical Office Building	Pelham	NH	8,399	259.65	New
	Podiatry Group	Marietta	GA	6,768	52.79	Renovation
	Marietta Podiatry Group	Marietta	GA	4,400	241.14	New

All prices are updated to January 1, 2019 and are national averages.
For a more in-depth report of any of these buildings or additional case studies contact
Design, Cost and Data at 800-533-5680, or go to www.DCD.com

2019 SQUARE FOOT TABLES

PROJECT	DESCRIPTION	CITY	STATE	SIZE	$/SF	NOTES
	Office					
Office	Restaurant Support Center	Lenexa	KS	186,465	255.30	New
	5000 NASA Boulevard	Fairmont	WY	132,000	272.23	New
	Rockford Construction Office	Grand Rapids	MI	71,144	108.77	Adaptive Reuse
	Woodlawn Office Bldg. (Shell Only)	Louisville	KY	60,000	137.60	New
	Rosecrance Ware Center	Rockford	IL	44,800	135.80	Adaptive Reuse
	Professional Center (Shell)	White Marsh	MD	43,025	173.11	New
	Swan Skyline Office Plaza (Shell)	Tucson	AZ	37,200	116.75	New
	Infinite Energy Phase IV	Gainesville	FL	36,500	306.05	New
	Freedom Plaza Building	Cookeville	TN	28,488	239.28	New
	Landmark Professional Building	Clayton	NC	27,231	226.88	New
	FC Gulf Freeway Building (Shell)	Houston	TX	24,084	258.29	New
	White Street Building (Shell Only)	Marietta	GA	23,809	210.78	New
	Columbia Shores Office Condo	Vancouver	WA	22,574	174.94	New
	Office & Design Studio	Chicago	IL	20,244	122.82	Tenant Build-out
	Pinnacle III Office Tenant Finish	Leawood	KS	18,409	58.47	Tenant Build-out
	Commerce Park (Shell)	Suwanee	GA	17,097	119.91	New
	PCWA Business Center Interior	Auburn	CA	12,085	60.82	Renovation
	Tenth Avenue Holdings Offices	N.Y.C.	NY	11,000	85.37	Tenant Build-out
	Office Park - Building A (Shell)	Fort Collins	CO	10,000	262.18	New
	Longshoremens Welfare Fund Building	Savannah	GA	8,160	384.93	New
	Garry Street Office Building	Manitoba	CA	7,506	153.59	Renovation
	Reserve Advisors	Milwaukee	WI	5,300	49.24	Tenant Build-out
	510 Armory Street Office	Boston	MA	5,100	94.28	Renovation
	Offices of Bonsall Shafferman	Bethlehem	PA	4,950	69.74	Tenant Build-out
	FCT Capital Partners	Houston	TX	4,100	81.58	Tenant Build-out
	Martin Rogers Associates Office	Wilkes-Barre	PA	4,000	130.54	Addition/Renovation
	Cowan & Kohne Financial	Suwanee	GA	3,713	122.93	Tenant Build-out
	212 Archer Street Office	Bel Air	MD	3,600	173.89	New
	Utilities Analyses Inc.	Suwanee	GA	3,513	114.02	Tenant Build-out
	Visual Lizard Interior Fit-Up	Manitoba	CA	2,430	91.49	Tenant Build-out
	Richardson State Farm	Houston	TX	2,200	110.62	Tenant Build-out
Mixed-Use	Office/Retail/Parking Mixed-Use	Jackson	MS	228,407	250.99	New
	Korte & Luitjohan Office & Shop	Highland	IL	26,000	120.60	New
Medical Office	Evanston Medical Office Building	Evanston	WY	9,157	276.44	New
	Advanced Medical Group	Suwanee	GA	4,433	117.38	Tenant Build-out
	Bothell Dental Office Build Out	Bothell	WA	2,203	194.92	Tenant Build-out
Headquarters	CONSUL Energy Corporation Headquarters	Southpointe, Canonsburg	PA	317,500	232.80	New
	Fairmont Supply Corporate Headquarters	Southpointe, Canonsburg	PA	75,255	153.85	New
	Practice Velocity Corporate Headquarters	Machesney Park	IL	64,318	96.67	Adaptive Reuse
	Enterprise Integration Headquarters	Jacksonville	FL	57,723	60.15	Renovation
	Linear Technology	Cary	NC	20,000	312.43	New
	PIPS Technology Inc.	Knoxville	TN	19,884	248.35	New
	Lee Steel Corporate Offices	Novi	MI	15,781	165.06	Renovation
	Weaver Cooke Headquarters	Goldsboro	NC	15,464	297.02	New
	In Capital Holdings	Boca Raton	FL	13,000	166.92	Tenant Build-out
	ACCION Regional Headquarters	Albuquerque	NM	7,580	319.55	New
Civic Office	Miss Department of Environmental Quality	Jackson	MS	121,170	93.09	Renovation
	County Central Office Complex	Pensacola	FL	74,630	282.13	New
	State of WV Office Building	Fairmont	WV	70,442	251.37	New
	JAX Chamber of Commerce Renovation	Jacksonville	FL	20,110	207.55	Renovation

2019 SQUARE FOOT TABLES

PROJECT	DESCRIPTION	CITY	STATE	SIZE	$/SF	NOTES
	Recreational					
Educational	The Pavilion at Ole Miss	Oxford	MS	235,301	441.85	New
	University Laker Turf Building	Allendale	MI	137,662	146.50	New
	CSU Recreation Center	Chico	CA	110,245	455.88	New
	Intramural Recreation Penn State	State College	PA	59,303	404.40	Addition/Renovation
	Center For Women's Athletics	Fayetteville	AR	39,183	360.11	New
	Pickens Recreation Center	Pickens	SC	20,400	179.76	New
	High School Concessions & Press Box	Loganville	GA	1,506	461.85	New
Health Club	Brooklyn Yard Fitness Club (Shell)	Portland	OR	63,987	113.17	New
	Title Boxing Club	Cedar Hill	TX	4,330	70.16	Tenant Build-out
Recreational	Community Recreation Center	Williston	ND	223,787	439.84	New
	Spirit Lake Casino & Resort	St. Michael	ND	112,277	59.66	Renovation
	Phipps Tropical Forest	Pittsburgh	PA	80,000	321.02	New
	Family Recreation Center	Colonie	NY	70,256	234.63	New
	Youth Activity Center	Joplin	MO	62,056	99.36	New
	New Holland Recreational Center	New Holland	PA	51,256	129.56	Renovation
	Community College Recreation Center	Cedar Rapids	IA	43,500	163.49	New
	Anderson Recreation Center	Anderson	SC	34,282	366.57	New
	East Park Community Center	Nashville	TN	33,000	294.05	New
	Church Family Life Center	Clemson	SC	31,509	317.31	New/Renovation
	C.K. Ray Recreation Center	Conroe	TX	30,380	149.75	Addition/Renovation
	St. Raphael Athletic & Wellness Center	Pawtucket	RI	30,268	259.27	New
	The Forge For Families	Houston	TX	29,860	244.45	New
	Christian Life Center	Birmingham	MI	26,966	334.59	Addition
	Children's Sports Center	Woodbury	MN	26,219	119.66	New
	Trinity River Audubon Center	Dallas	TX	20,791	922.63	New
	Baptist Church Activity Center	Indianapolis	IN	16,636	147.94	New
	Boys & Girls Club Syracuse	Syracuse	NY	12,107	180.81	Addition
	Job Corp Recreational Building	Hartford	CT	11,300	202.88	New
	Presbyterian Family Life Center	Strawberry Plains	TN	11,236	164.85	New
	McDaniel Yacht Basin	North East	MD	7,620	185.67	New
	Bicentennial Park	Cincinnati	OH	4,050	869.74	New
	Bahosky Softball Complex	Bronx	NY	3,800	470.35	New
Swimming Center	Resort & Indoor Waterpark	Cortland	NY	175,060	262.76	New
	The Aquatic Center	Tunica	MS	45,008	313.09	New
	Spirit Lake Phase 4	St. Michael	ND	26,630	357.96	Addition/Renovation
	Family Aquatic Center	Beachwood	OH	7,500	1,163.31	New
	Community Aquatic Park & Center	Billings	MT	6,730	731.64	New
Theater	Cinema & IMAX Theatre	Lansing	MI	13,750	221.16	New
	Academy Theater	N.Y.C.	NY	4,593	224.69	Renovation
YMCA	David D. Hunting YMCA	Grand Rapids	MI	162,966	201.54	New
	YMCA Recreational Center	Ann Arbor	MI	83,377	274.31	New
	Floyd Co. YMCA & Aquatic Center	New Albany	IN	82,324	334.00	New
	Wade Walker Park Family YMCA	Stone Mountain	GA	59,134	348.62	New
	Alexandria YMCA	Alexandria	MN	55,150	168.68	New
	Greater Nashua YMCA	Nashua	NH	49,980	207.62	New
	Lancaster YMCA Harrisburg Ave.	Lancaster	PA	42,502	353.07	New
	Greater Kingsport Family YMCA	Kingsport	TN	40,007	252.20	New
	Eastside YMCA	Knoxville	TN	39,984	250.57	New
	Highland County Family YMCA	Hillsboro	OH	33,228	149.29	New
	Houston Texans YMCA	Houston	TX	31,628	354.28	New
	Cypress Creek YMCA	Houston	TX	25,699	149.42	Addition/Renovation

All prices are updated to January 1, 2019 and are national averages.
For a more in-depth report of any of these buildings or additional case studies contact
Design, Cost and Data at 800-533-5680, or go to www.DCD.com

2019 SQUARE FOOT TABLES

PROJECT	DESCRIPTION	CITY	STATE	SIZE	$/SF	NOTES
		Religious				
Church	First United Methodist Church	Orlando	FL	121,536	278.88	New
	Beautiful Savior Lutheran Church	Plymouth	MN	69,700	114.06	New
	Solid Rock Baptist Church	Berlin	NJ	58,359	103.13	New
	North Side Baptist Church	Greenville	SC	48,087	263.36	New
	St. Martha Catholic Church	Porter	TX	46,748	520.37	New
	Gracepoint Gospel Fellowship Church	Ramapo	NY	46,595	193.55	New
	Good Shepherd Methodist Church	Odessa	TX	41,003	308.27	New
	Davisville Church Addition	Southampton	PA	36,090	182.25	Addition
	Immaculate Catholic Church	Columbia	IL	34,000	254.85	New
	Keystone Community Church	Ada	MI	29,775	150.06	New
	Grace Church	Des Moines	IA	29,296	203.60	New
	Good Shepherd Church	Naperville	IL	27,869	201.68	Addition/Renovation
	River Hills Baptist Church	Corpus Christi	TX	27,404	305.32	New
	Good Shepherd Catholic Church	Smithville	MO	24,810	209.75	New
	Prince of Peace Catholic Church	Chesapeake	VA	24,740	262.81	Addition/Renovation
	Sanctuary Addition Christian Church	Oklahoma City	OK	23,820	157.53	Addition
	St. Sylvester Catholic Church	Gulf Breeze	FL	22,000	413.07	New
	St. Peters Catholic Sanctuary	Fallbrook	CA	20,764	455.15	New
	Notre Dame Catholic Church	Houston	TX	20,280	419.88	New
	St Eugene Catholic Church	Oklahoma City	OK	20,000	460.22	New
	Chapin Presbyterian Church	Chapin	SC	19,900	335.55	New
	St. Michaels Catholic Church	Glen Allen	VA	19,770	360.00	New
	Shrine of Holy Spirit	Branson	MO	19,200	342.59	New
	Wildwood United Methodist Church	Magnolia	TX	19,000	244.79	New
	Episcopal Church of the Nativity	Scottsdale	AZ	18,288	122.43	Adaptive Reuse
	Hardin Church of Christ	Knoxville	TN	17,149	123.08	New
	First United Methodist Church	Crossville	TN	15,816	481.73	New
	Good Shepherd Episcopal Church	Silver Spring	MD	15,200	287.43	Addition/Renovation
	Ascension Catholic Church	LaPlace	LA	15,057	379.42	New
	St. Patrick Catholic Church	Jacksonville	FL	14,139	271.21	New
	Our Lady of Guadalupe Catholic Church	Rosenberg	TX	12,910	418.31	New
	St. Timothy's Episcopal Church	Creve Coeur	MO	12,682	208.54	New/Renovation
	St. Paul Lutheran Church	Pomaria	SC	12,072	341.44	New
	Covenant Baptist Church	Florida City	FL	10,725	254.76	New
	First United Methodist Church	Katy	TX	10,503	309.43	Addition/Renovation
	Lake Ann United Methodist Church	Lake Ann	MI	9,975	213.82	New
	United Methodist Church	Odenton	MD	8,783	334.09	New
	Kent R. Hance Chapel at Texas Tech	Lubbock	TX	6,530	606.44	New
	Haven for Hope Chapel	San Antonio	TX	2,232	455.50	New
Multi-Purpose	Baptist Church Multi-Purpose Bldg	Maryville	TN	41,656	113.65	New
	Good Shepherd Parish Center	San Diego	CA	28,752	156.75	New
	Christian Life Center	Kansas City	MO	26,320	304.46	New
	Baptist Church Outreach Center	Fort Smith	AR	25,000	267.32	New
	St Rafael Administration Building	San Diego	CA	24,276	168.83	New
	Catholic Church Social Hall	Chula Vista	CA	23,596	280.91	New
	United Methodist Church	West Chester	PA	11,935	245.65	Addition/Renovation
	Student Ministry Center	Knoxville	TN	11,700	304.33	New
	Catholic Pastoral Ministries Center	Spring	TX	10,135	390.73	New
	Christian Renewal Center	Dickinson	TX	8,500	214.80	New
	Holy Family St Lawrence Parish Center	Essex Junction	VT	7,900	225.27	New
	Presbyterian Church Addition	Gap	PA	7,414	258.60	Addition/Renovation
	New Hope Church Addition/Alteration	Saint Louis	MO	5,564	198.00	Renovation

All prices are updated to January 1, 2019 and are national averages.
For a more in-depth report of any of these buildings or additional case studies contact
Design, Cost and Data at 800-533-5680, or go to www.DCD.com

SF-11

2019 SQUARE FOOT TABLES

PROJECT	DESCRIPTION	CITY	STATE	SIZE	$/SF	NOTES

Residential

	PROJECT	CITY	STATE	SIZE	$/SF	NOTES
Apartment	Solace Apartments	Virginia Beach	VA	331,681	100.23	New
	1221 Broadway Lofts	San Antonio	TX	205,137	141.56	Adaptive Reuse
	Sustainable Fellwood Phase I	Savannah	GA	124,037	132.42	New
	Kelly Cullen Community	San Francisco	CA	98,385	548.70	Adaptive Reuse
	Bachelors Enlisted Quarters	Camp Williams	UT	76,253	249.49	New
	Mockingbird Terrace Homes	Louisville	KY	71,110	158.57	New
	Homeless Men's Residential	San Antonio	TX	67,908	262.84	New
	Homeless Women's/Family Residence	San Antonio	TX	60,182	253.27	New
	Magnolia Place	Lancaster	PA	39,714	159.73	New
	Elkins First Ward Apartments	Elkins	WV	27,000	114.67	Adaptive Reuse
	Young Burlington Apartments	Los Angeles	CA	24,399	205.49	New
	The Lofts at 300 Bowman	Dickson City	PA	23,900	85.67	Adaptive Reuse
	Peaceful Paths Emergency Svc Campus	Gainesville	FL	22,535	147.04	New
	Wylie House - Ronald McDonald House	Kansas City	MO	21,885	186.50	New
	Dogwood Manor Apartments	Oak Ridge	TN	19,975	169.16	New
	Anderson Village Multi-Family	Austin	TX	12,500	282.39	New
	Salvation Army Sally's House	Houston	TX	7,812	246.44	Addition
	Stones River Apartment Complex	Murfreesboro	TN	7,548	227.89	Addition
	Sunshine Park Apartments Renovation	Gainesville	FL	2,252	120.79	Renovation
Assisted Living	Kenmore Apartments Senior Housing	Chicago	IL	90,528	211.29	Renovation
	Country Meadows Retirement	Allentown	PA	53,237	161.68	New
	Creekside Village Assisted Living	Harrisburg	PA	16,150	140.51	New
	Landis Homes Retirement Community	Lititz	PA	14,255	66.41	Renovation
Dormitory	NSU Graduate Student Housing	Davie	FL	203,500	217.08	Renovation
	Rider University Student Housing	Lawrenceville	NJ	50,500	233.50	New
	JWU Biscayne Commons Dormitory	Miami	FL	40,048	256.13	New
	College Residence Dorm	Bloomfield	NJ	25,980	315.37	Renovation
Single-Family Home	Island Residence	Grosse Ile	MI	19,237	680.75	New
	MG Residence Restoration	Williamston	MI	9,768	61.23	Renovation
	Concepcion House	Coral Gables	FL	6,067	351.09	New
	Leal House	Miami	FL	5,935	242.55	New/Renovation
	Monserrate Street Residence	Coral Gables	FL	5,885	438.67	New
	Private Residence	Newburgh	IN	5,566	349.72	New
	Private Residence	Austin	MN	5,489	144.09	New
	Private Residence	Lake Wallenpaupack	PA	4,845	315.56	New
	Island in the Grove	Boca Raton	FL	4,701	360.73	New
	Private Residence	Benson	AZ	3,660	175.95	New
	Fairhope Green Home	Fairhope	AL	3,610	200.32	New
	PATH Concept House	Omaha	NE	3,490	76.74	New
	Private Residence	La Jolla	CA	3,420	354.09	New
	Solar House - Private Residence	Fly Creek	NY	3,304	177.21	New
	Elliott Residence	Fort Collins	CO	3,300	267.23	New
	Renfrew House	Manitoba	CA	3,206	191.24	New
	Rosado I Hansen Residence	Tucson	AZ	3,175	133.40	New
	306 W. Waldburg Residence	Savannah	GA	2,588	167.28	New
	Nutter Green Home	Milford	OH	2,289	159.52	New
	Guest House Residence	Ahwatuckee	AZ	1,913	389.38	New
	Kiwi House	Baton Rouge	LA	1,515	155.50	New
	Private Residence Renovation	Shavertown	PA	810	161.64	Renovation

All prices are updated to January 1, 2019 and are national averages.
For a more in-depth report of any of these buildings or additional case studies contact
Design, Cost and Data at 800-533-5680, or go to www.DCD.com

UNITED STATES INFLATION RATES
July 1, 1970 to June 30, 2018
AC&E PUBLISHING CO. INDEX

Start July 1

YEAR	AVERAGE %	LABOR %	MATERIAL %	CUMULATIVE %
1970-71	8	8	5	0
1971-72	11	15	9	11
1972-73	7	5	10	18
1973-74	13	9	17	31
1974-75	13	10	16	44
1975-76	8	7	15	52
1976-77	10	5	15	62
1977-78	10	5	13	72
1978-79	7	6	8	79
1979-80	10	5	13	89
1980-81	10	11	9	99
1981-82	10	11	10	109
1982-83	5	9	3	114
1983-84	4	5	3	118
1984-85	3	3	3	121
1985-86	3	2	3	124
1986-87	3	3	3	127
1987-88	3	3	4	130
1988-89	4	3	4	134
1989-90	3	3	3	137
1990-91	3	3	4	140
1991-92	4	4	4	144
1992-93	4	3	5	148
1993-94	4	3	5	152
1994-95	4	3	6	156
1995-96	4	3	6	160
1996-97	3	3	3	163
1997-98	4	3	4	167
1998-99	3	3	2	170
1999-2000	3	2	3	173
2000-2001	4	4	4	177
2001-2002	3	4	3	180
2002-2003	4	5	3	184
2003-2004	3	5	2	187
2004-2005	8	4	10	195
2005-2006	8	6	9	203
2006-2007	8	4	11	212
2007-2008	3	4	3	215
2008-2009	4	3	5	219
2009-2010	3	5	1	222
2010-2011	2	3	2	224
2011-2012	3	3	3	227
2012-2013	3	3.5	3	230
2013-2014	2	1.8	2	232
2014-2015	1.4	1.8	1	233.4
2015-2016	1	1.8	0	234.4
2016-2017	1	1.8	0	235.4
2017-2018	5	4.5	5	240.4
2018-2019	3.5	2.3	4.6	244

Design Cost Data™ DCD

INDEX

INDEX

INDEX

INDEX